MATHEMATICS FOR THE
MILDLY HANDICAPPED

MATHEMATICS FOR THE MILDLY HANDICAPPED
A Guide to Curriculum and Instruction

JOHN F. CAWLEY
State University of New York at Buffalo

ANNE MARIE FITZMAURICE-HAYES
University of Hartford

ROBERT A. SHAW
University of Connecticut

ALLYN AND BACON, INC.
Boston   London   Sydney   Toronto

Library of Congress Cataloging-in-Publication Data

Cawley, John F.
Mathematics for the mildly handicapped.

Bibliography: p. 246
Includes index.
1. Mathematics—Study and teaching (Elementary)
2. Handicapped children—Education—Mathematics.
I. Fitzmaurice-Hayes, Anne Marie. II. Shaw, Robert A.
III. Title.
QA135.5.C383 1988   371.9'044   87-19317
ISBN 0-205-11081-9

Printed in the United States of America

10 9 8 7 6 5 4 3 2 1 92 91 90 89 88 87

CONTENTS

Contents

PREFACE

This book is about curriculum and instruction in mathematics for mildly handicapped children. It is designed to highlight two phenomena. One phenomenon is mathematics; the other is mildly handicapped children. It is our belief that the needs of mildly handicapped children and the content of mathematics can be mutually enhancing when both are viewed from a comprehensive perspective.

An adventure into the world of mathematics is endless. The variations in both the characteristics and needs of mildly handicapped children are extensive. Both the range of mathematics and the descriptions of mildly handicapped children are selective and represent our own backgrounds and research in these topics. The mathematics is restricted to topics that we have found appropriate for and attainable by mildly handicapped children. This does not mean that all children performed perfectly in all topics; it does mean that some children performed in all areas and that all children performed in some areas. Our evidence for this statement stems from our evaluation of the performance of the individuals who were under consideration during the development of *Project MATH* (Cawley et al. 1974, 1976) and *Multi-modal-mathematics* (Cawley et al. 1980). The evidence included appraisals by every instructional guide and teacher involved in these projects. Not all children mastered the different concepts and skills at the same rate. Some attained mastery in a few days or a few weeks; others took longer. We learned that both the instructional needs and the curriculum needs of mildly handicapped children require equivalent, but not necessarily equal, consideration.

Our priority is curriculum. We believe that decisions as to what to teach, when to teach it, and for how long are more crucial than how to teach it. The eight mathematics chapters in this book are sequenced in a manner different from many other sources. We begin with space, spatial relations, and figures, the study of which is more commonly known as geometry; we have found this to be among the easiest areas of mathematics for mildly handicapped children and among the most essential in real life. We follow with problem solving. Even though we believe that problem solving should be introduced by kindergarten and should be the reason for computation rather than an afterthought, we end with it in this book because our curriculum emphasis recognizes that mildly handicapped children will be educated in a variety of regular classes and special education units of service delivery. Special education for the mildly handicapped child cannot exist on a curriculum limited to arithmetical computation. Full service necessitates comprehensive curriculum planning.

The book begins with two chapters that contain a discussion of selected characteristics of mildly handicapped children and the meaning of these characteristics for certain educational practices. Also included are a review and analysis of learning and instruction as related to mathematics and mildly handicapped children.

The eight mathematics chapters present the material we consider both necessary and realistic in long-term comprehensive programming for mildly handicapped children. It is important to note that we consider what we present to be a minimum for most, if not all, mildly handicapped children. Considerable emphasis is placed on concepts and utilization. We acknowledge the importance of computation and the need for mildly handicapped children to master it, but we feel that a program rooted primarily in computation does not meet the lifelong needs of the mildly handicapped child.

Chapters on assessment and multimedia programming conclude the text. The latter describes the use of a variety of media in teaching mathematics to the mildly handicapped and concludes with a set of examples for using a microcomputer.

Our primary goal for this book is to provide a basis for qualitative approaches to mathematics for the mildly handicapped. We hope that it serves as a guide in the search for an optimum education for the mildly handicapped.

John F. Cawley
Anne Marie Fitzmaurice-Hayes
Robert A. Shaw

CHAPTER ONE

MILDLY HANDICAPPED CHILDREN

This is a book about mathematics for mildly handicapped children. Our intent is to present an approach to mathematics that will enable regular class and special education teachers to stress both *concept* and *skill* development to meet the needs of mildly handicapped children in mathematics. Considerable emphasis is placed on problem solving and the contributions of problem solving to the overall development of mildly handicapped children. We believe that mathematics programs for the mildly handicapped must serve two purposes. First, they must meet the learner's *qualitative* need for mathematics. Second, mathematics programs must be designed to contribute to the overall development of the mildly handicapped in areas such as language, cognition, and social-emotional needs.

We have therefore made certain decisions relative to content or curriculum priorities. For the most part, the mathematics topics resemble those found in regular education curricula more than those found in special education curricula. The reason for this is that special education curricula tend to stress arithmetic computation, forsaking important topics in geometry, measurement, and problem solving, whereas regular education curricula tend to include a variety of topics. Further, about 80 percent of the learning disabled and about 40 percent of the mildly retarded receive the dominant portion of their mathematics education in regular classes. It is therefore the responsibility of both the regular educator and the special educator to provide the mildly handicapped child with comprehensive experiences in mathematics.

We have excluded reference to certain topics whose primary virtue rests in their value to mathematics for the sake of mathematics. Prime numbers is one such topic. Our reason for this decision is that those mildly handicapped children who can master topics such as prime numbers are not children who require curriculum modifications; they are more likely to require instructional modifications. Modifications in instruction require consistent collaborative efforts between special educators and regular educators.

In the mathematics chapters in this book, we stress concepts over skills, because many concepts are understood by young children long before they can perform skills. For example, young children deal quite effectively with the division of a whole number by a fraction in real situations, such as cutting an apple into halves or near halves, even though these children have no idea of the computational algorithm.

The approach in this book has its roots in a comment by Seguin (1849) in his discussion of teaching the mentally retarded. Seguin's comment, "Better one thing understood than one hundred things remembered," has long been a standard in our approach to mathematics for the mildly handicapped. Any program of curricula and instruction predicated on such a statement must emphasize meanings and understanding in preference to rote computation. Speed and habituation are essential considerations in mathematics programming for the mildly handicapped, but these should come after meanings and understanding, not in place of them.

PLANNING FOR THE MILDLY HANDICAPPED

For purposes of this book, the term *mildly handicapped children* will refer to children whose developmental characteristics indicate that they will be capable of a reasonable degree of independent living as adults. In this book, the term includes children who are legally identified as (1) *mildly retarded,* (2) *learning-disabled,* and (3) *behaviorally disordered.*

Mildly handicapped children are therefore children who can derive significant benefits from the academic, vocational, and personal-social offerings of school. For some of these children, the thrust of the school program will have a curriculum orientation. For others the program thrust will be instructional. In other cases, the school program should accentuate social-personal development. For still others, the program will consist of a combination of curriculum, instructional, and personal-social offerings.

Mathematics is defined as a reasonable representation of the concepts and skills of numeration, numbers, fractions, geometry, measurement, and problem solving. We have limited the contents of this book to a "reasonable representation" of mathematics because the developmental characteristics of many of the children about whom this book is written limit the amount of math that falls within their capabilities. It is important, therefore, that we distinguish those children who are capable of doing more math, but are not doing it, from those children who are entered in math programs they are not able to manage. It is also important that we define the mathematics curriculum in terms of its content, not its grade level. As an example, it has never been shown that the cognitive development of the mildly retarded is sufficient for them to accomplish the mathematics of the secondary school. True, they might do fine in a secondary school course that emphasizes social applications of the mathematics of the elementary grades. But this is not secondary school mathematics. Therefore, the primary emphasis in programming for the mildly retarded ought to be in curriculum, and this fact should be recognized.

By contrast, some learning-disabled children are quite capable of doing secondary school mathematics if the instructional program is adapted to meet their unique acquisition-and-response needs (Shaw 1985). For these children, the program emphasis would be on instruction. Emphasis on curriculum would be appropriate for those learning-disabled children whose developmental characteristics suggest long-term cumulative deficits in mathematics. The primary curriculum decision by special educators relative to this latter group typically stresses proficiency in whole-number computation.

The regular class teacher, on the other hand, is often required to present a variety of mathematics topics over a given period. This means that for those learning-disabled children who experience difficulty with mathematics, one must decide how much time should be devoted to the development of competence in the operations on whole numbers at the expense of experiences with other topics. This is an important consideration for present-day mathematics programming because of the availability of calculators, microcomputers, and other devices that facilitate computational performance.

The third group of children, the behaviorally disordered, may or may not have difficulties in mathematics. The school problem with these children may focus more on their unwillingness to do the assignments and on the irregularity with which they may do them than on their competence. For these children, it may take only a mathematics orientation that is interesting, satisfying, and rewarding to get them to perform at their level of ability. When a specific difficulty exists in mathematics, it may be necessary to conduct a comprehensive appraisal and implement a program of support services.

CLASSIFICATION

The Use of Labels

We raise the issue of labels and their use because we would like to provide the regular class teacher and the special class teacher with some knowledge of these concerns within the field of education.

Three legally identifiable groups of children make up the population of mildly handicapped children for whom this book is written. As mentioned earlier, these are the *mildly retarded,* the *learning disabled,* and a third group referred to as the *seriously emotionally disturbed* or *behaviorally disordered.*

Data show there are approximately 4,100,000 handicapped children being served in the United States. About 2,900,000 of these children receive the dominant part of their education in regular classes (*Executive Summary* 1984). Approximately 40 percent of the handicapped are regarded as learning disabled, about 15 percent are mildly retarded, and some 8 percent are seriously emotionally disturbed or behaviorally disordered. These three groups constitute about 60 percent of the population identified as handicapped children and nearly all of those who are reported to be mildly handicapped. These are important data for regular educators, who provide the great majority of the mathematics education for the mildly handicapped.

Although the matter is largely philosophical, any approach to the education of mildly handicapped children must deal with the issue of labels, because state education departments, local school districts, and units of higher education differ in their views of the matter. Some argue that labels and the systems of classification that accompany them have little or no functional use for school programming. Others find labels useful. Our perspective is that the best way lies in between the use of labels and their abandonment.

If someone were to initiate a plan to develop a school math program for six-year-old mildly retarded children, this would suggest a line of thinking somewhat different from that in developing a school math program for learning-disabled or behaviorally disordered children. The most immediate thought that occurs relates to curriculum planning. Mildly retarded children, although manifesting much of the heterogeneity of other children, are characterized by reasonably stable patterns of development (Zigler, Balla, and Hodapp 1984).

This contrasts with the heterogeneous, but unstable, patterns of development among learning-disabled children (Allardice and Ginsburg 1983). The patterns of development among behaviorally disordered children are also heterogeneous and unstable, with major variation in age of onset; such disorders may occur at any developmental level.

For the mildly retarded, long-term programming should take place in a curriculum perspective. For the learning disabled or behaviorally disordered, the curriculum decisions might be either short-term or long-term. They may relate only to specific deficits or needs and may not encompass the broad curriculum planning that is appropriate for the mildly retarded.

Behaviorally disordered children may exhibit transient or chronic behavior patterns, evidenced at one point in time but not another. They may be situational, occurring in one classroom but not in another. They may be cause-specific (e.g., family divorce). The line of thinking stimulated by the term "behaviorally disordered" suggests further appraisal, specifically in terms of environmental variables, personal qualities, and academics. Should academic deficiencies be ruled out as a source of difficulty, programming would take on a mental health approach, which could be behaviorally or "therapeutically" defined. It may be possible, for example, for the regular class teacher to modify the environment of the classroom to meet the needs of some of these children.

Individuality

The use of labels to describe individual children has little merit. Individual programming is a function of performance characteristics in academic and social-personal domains. Children either perform school work or they do not; they either demonstrate the behaviors expected at school or they do not. It is the set of academic and social-personal attributes set forth by the school, and the extent to which a child meets or fails to meet all or some of the criteria relative to these attributes, that brings about the referral for special education services. These same combinations of factors lead to assignment to service delivery units. In fact, in some schools, combinations of academic and social factors control the extent to which the child is educated along with other children (Truesdell 1985). Thus, it is the set of individual learner attributes, as they are viewed within the context of school, that provides the basis for individual programming. Children differ, and within any group of children there are sets of children who differ from one another.

Subtyping

Ongoing research in learning disabilities (Lyon 1985) suggests that there are subtypes of learning-disabled children, and there is the possibility that some instructional procedures or curriculum options are more appropriate for some learning-disabled children than for others. Subtyping was the precursor to the field of learning disabilities, when Strauss and Lehtinen (1947) differentiated mentally retarded children in terms of brain-injured (exogenous) and non-brain-injured (endogenous) groups.

Extensive subtyping has been reported for many years among children with reading disabilities and among children with mathematics disabilities (Cawley et al. 1979a; Kosc, 1974).

In mathematics, it has been demonstrated that there are subtypes among the learning disabled (Ackerman, Anhalt, and Dykman 1986; Cawley et al. 1979a). Samples of mildly retarded children, for example, have been differentiated on single factors, such as arithmetical problem solving (Sedlak 1973).

We credit Kosc (1974) with directing our attention to the existence of subtypes in mathematics among children with learning problems. Kosc defined six types of dyscalculia:

- Verbal dyscalculia, which is manifested by disturbed ability to deal verbally with mathematical terms and relations
- Practognostic dyscalculia, in which there is a disturbance in the manipulation of real or pictured objects to represent mathematical ideas

- Lexical dyscalculia, which is a disturbance in reading mathematical symbols
- Graphical dyscalculia, in which the individual has a disability in writing mathematical terms or symbols
- Ideognostical dyscalculia, where there is a disturbance in understanding mathematical meanings and relations
- Operational dyscalculia, in which the individual has a weakness in carrying out the operations of arithmetic

Ackerman, Anhalt, and Dykman (1986) examined arithmetical automatization failure in subtypes of children with or without attention disorder deficits. Contrasts were made among four groups as follows:

| | *Age in months* | |
	Younger	*Older*
Control	109	131
Hyperactive with attentional disorder	108	129
Developmental reading disorder without hyperactivity	114	134
Attention disorder deficit with normal reading	109	129

Selected findings show:

No differences in WRAT Arithmetic performance among the four groups.

Boys with adverse teacher ratings on arithmetic performance were not substantially different from boys not so rated, as evidenced by test data that failed to confirm teacher judgment in some 40 percent of the cases.

On five different arithmetic work samples, there were no differences among subjects or among work samples.

Boys with reading disabilities or attentional disorder deficits were more commonly found to be slow and inaccurate in computational accuracy and speed.

Subtyping was possible. Subjects with WRAT Arithmetic scores at or below the 25th percentile were differentiated into three groups:

One group had reading scores higher than arithmetic scores.

One group had equivalent reading and arithmetic deficiencies.

One group showed deficits in all areas of performance.

Most children diagnosed as having early reading disability will later manifest equivalently poor arithmetic performance.

Children in this study with attentional disorder deficits evidenced major difficulties in memory in contrast to spatial reasoning.

Short-term memory difficulties cross all categories of mildly handicapped children. Webster (1980) studied short-term memory among mathematically proficient and mathematics-deficient children as a function of input-output modality pairings. The stimuli consisted of strings of visually presented consonants and aurally presented digits. Mathematically proficient children (MP) achieved at the expected sixth-grade level. One sample of deficient children

TABLE 1.1 Percent of information lost from immediate recall to short-term (ST) and long-term (LT) recall by learning group

Learning group	Visual ordered		Visual unordered		Auditory ordered		Auditory unordered	
	ST	*LT*	*ST*	*LT*	*ST*	*LT*	*ST*	*LT*
Average	20.5	24.2	13.1	29.0	23.3	29.9	21.0	26.5
Emotionally disturbed	20.1	23.5	17.6	22.1	26.0	31.8	25.7	27.8
Learning-disabled	30.4	34.0	25.5	30.4	40.8	51.6	35.5	47.9
Mildly re-tarded	27.3	53.0	24.5	39.6	41.5	51.0	48.5	53.0

Source: R. E. Webster, *Learning Efficiency Test.* Novato, CA: Academic Therapy Publications, 1981. Reprinted with permission of the author.

(MD1) were one to one-and-one-half years delayed, and another group (MD2) was more than two years delayed. Mathematically proficient children performed significantly better than the MD2 sample. There were no other differences among samples. Input data showed that visual input was better than auditory input. Output data showed MP better for visual than auditory. There were no differences between input and output pairings for MD1, whereas MD2 children showed preferences for auditory output as opposed to visual output.

Webster (1981) also calculated the percent of loss for immediate recall (a few seconds), short-term recall (about twenty seconds with an interpolated counting task) and long-term recall (about twenty seconds later with an interpolated sentence repetition task) among nonhandicapped children and three groups of mildly handicapped children. His data are shown in Table 1.1.

As can be seen, percent of loss under each condition was different for the four groups. Percent of loss ranged from near 20 percent for nonhandicapped children to better than 50 percent under some conditions for mildly retarded and learning-disabled children. Correlations between LET (Learning Efficiency Test) scores and reading and arithmetic achievement range from .49 for average children and reading to .93 for mildly retarded children in reading and math.

In contrast to the above, where different types of data are collected on the developmental characteristics of children, direct-instruction advocates, such as Blankenship (1984, 1985) and Blankenship and Lovitt (1976), collect data on the immediate problem of the child and proceed to intervene on that specific problem. If, for example, the difficulty is in subtraction, the teacher will conduct an analysis of the child's performance in subtraction and develop a program of intervention for subtraction (Blankenship 1982). If the difficulty is in problem solving, direct instruction would be provided (Blankenship and Lovitt 1976). Performance, not labels, would guide the activities of direct-instruction advocates.

DEVELOPMENTAL CHARACTERISTICS

The sections that follow will describe certain developmental characteristics of our three target samples. Similarities and differences will be highlighted, and specific information relative to mathematics will be incorporated into the discussion.

Norman and Zigmond (1980) studied the characteristics of 1966 learning-disabled (LD) children aged 6.0–17.9. Mean reading grade equivalents were 1.8 for 6-year-olds to 5.75 for 17-year-olds. Math performance ranged from 1.23 for 6-year-olds to 6.77 for 17-year-olds. They note that achievement data reflect a lack of consistency when labeling students as LD. Younger children tended to perform at grade level. Achievement discrepancies of two or more years were not noted until twelve years of age. Taking a bit of editorial liberty, we interpret twelve years of age to approximate seventh grade in school. These data suggest, then, that the mean achievement level of the LD sample approximates fifth-grade level. This indicates that in the remaining five or six years of school, only one or two years of grade-equivalent gains are made. A continuing pattern tends to show that LD children make much of their progress in the first five to seven years of school. Little progress seems to be made beyond that level. This is a pattern that is strikingly similar to that discovered by Warner et al. (1980), in that their samples of learning-disabled children made little or no progress in reading, language, or mathematics between seventh and twelfth grades.

The pattern of research in mathematics among the mildly handicapped shows numerous efforts in computation, including work by Thornton and Toohey (1985) and Fleischner, Garnett, and Shepherd (1982), with little attention to problem solving (Fleischner and O'Loughlin 1985) or other mathematics topics with the learning disabled. However, research concerning the mildly retarded has many references to problem solving, such as studies by Cruickshank (1948), and by Cawley and Goodman (1969), as well as applications (Levine and Langress 1985).

Thornton and Toohey (1985) conducted a number of research projects on the teaching and learning of basic math facts among learning-disabled children in the United States and in Australia. Their results indicate that explicit intervention for fact recall is beneficial to LD students. In one of their studies, LD students used a special program of addition facts for 11 weeks, while regular class children continued with their prescribed program. At all grade levels (2–6), the program resulted in reducing the discrepancy between LD and regular class children. Note, however, that LD students were given 11 weeks of instruction on addition, whereas regular class children may have experienced a number of different mathematics topics.

In another part of their research (Thornton and Toohey 1985), LD children were taught basic addition facts for a 7-month period. Comparisons between pre- and post-test over a 7-month period showed raw score increases from 18 to 33 correct for third- and fourth-grade children and from 25 to 39 correct for fifth- and sixth-grade children. Post-test performance was probably underestimated because the children were given three minutes to complete the pretest and only two minutes to complete the post-test.

Performance of LD students was assessed in multiplication and division (Thornton and Toohey, 1985) in a program in which 27 teachers of the learning disabled taught all the math to their children while another 10 teachers taught math as a supplement to what the children were receiving in the regular class. Performance increased over the 9-month period by about the same amount, an increase of 25 items correct, for both multiplication and division for both teaching conditions.

Growth Patterns

Figure 1.1 displays simulated growth patterns in mathematics for four different developmental groups. These are included to stress the need for differentiated curriculum considerations by both the regular educator and the special educator.

The group labeled *above average* shows superior performance in mathematics status for all age levels. It is possible for this group to include children who are learning disabled or behaviorally disordered. These would be children whose academic disabilities are in areas other than mathematics. This pattern would clearly suggest that the learning disabilities manifest in the

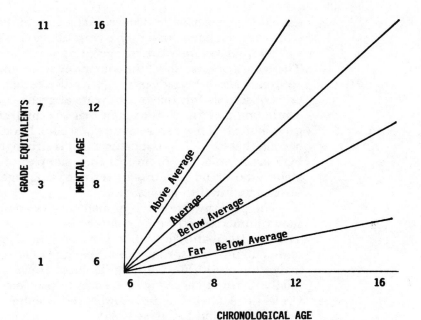

FIGURE 1.1 Simulated growth patterns

child are unrelated to mathematics or that any negative impact has been eliminated by instructional adaptations and coping skills (Alley and Deshler 1979).

There may be members of this group whose school marks are poor because they fail to do the required work or to meet school or teacher standards. However, when they do respond to school achievement or individual tests, their performance might reflect advanced standing. These children often fail in mathematics for reasons other than mathematics difficulties.

Mathematics expectancies for the advanced group would be consistent with the traditional college-preparatory program or the high-technology program of selected vocational-technical programs. Shaw (1985) has described adaptations in the college preparatory program for children whose mathematics ability and achievement is consistent with college preparatory standards but who are experiencing difficulty in math due to other factors.

Another group is termed *average*. This group is characterized by the traditional growth pattern, showing about one year of school achievement for each year of chronological growth. This group could include learning-disabled or behaviorally disordered children. In this instance, the performance of the child in mathematics would be similar to nonhandicapped children of the same age. This would occur when the effects of the disability are such that they do not impair performance in mathematics or when the effects of a disability have been eliminated by instructional and material modifications.

Particular attention needs to be directed toward mildly handicapped children who could be achieving at advanced or average levels in mathematics if the proper instructional and material adaptations were implemented. We express special concern for the child who lacks the reading skills needed to interpret and comprehend mathematics but who could otherwise perform satisfactorily in math. The reading disability handicaps the child receptively in that the child cannot access the message. The child then has no means of processing the information and making decisions on it, so his or her output is lacking or incorrect. Instructional activities that exchange messages through the active process of manipulation, the static process of fixed displays such as pictures, or the interactive exchange of oral language are all means by which equivalent representations of mathematics concepts can be exchanged.

The mathematics expectancies for this average group would include grade-level attainment at an appropriate rate throughout the elementary school. Number proficiency through long division, a capability with the operations of fractions and decimals (with the possibility there may not be full mastery of the three algorithms of percent), and competence in geometry and measurement are expected. Some children may need an additional year of basic mathematics before pursuing Algebra 1, and there is the likelihood that some will not go on to traditional courses in Algebra 2 or Geometry.

By definition, no mildly retarded children would have membership in the above average or average groups. Mildly retarded children tend to experience difficulty with complex multiplication and division. They perform certain fractions tasks, but, unfortunately, they do so with little understanding. More importantly, their performance is slow and laborious rather than rapid and quick. Our opinion is that many of the qualitative problems of the mildly retarded stem from matters that are as inherent in curriculum and instruction as they are intrinsic to the learner.

The term *below average* refers to a group for whom the developmental pattern shows clearly that grade-level achievement in mathematics is below chronological age expectancy. When a child of average or above-average cognitive ability is performing at this level, it is likely that the child manifests a form of learning disability or behavioral difficulty. The key diagnostic requirement is to determine the extent to which the problem exists as an overall developmental pattern or whether it is due to a single factor such as a mathematics difficulty, a reading deficiency, or some other specific difficulty. Should the child demonstrate an overall disability pattern, both curricula and instructional adaptations need consideration. Should the child demonstrate only a single area of deficit, it is possible that only instructional and material modifications are necessary.

The developmental pattern of the *below average* group is consistent with the developmental pattern of children who are mildly retarded. In this instance we have about two years of academic achievement for every three years of increase in chronological growth. In reality, overall group performance may not be as much as two years of achievement for three years of age growth. Three factors stand out when the growth pattern is viewed in relation to the mildly retarded. To begin with, these children may be eight or nine years of age before they are ready for formal instruction in reading and mathematics; during the first two or three years of school, such a child is functioning at a readiness level. The mathematics of those years should be no more complex than what an average child experienced at four or five years of age. In our own work in *Project MATH* (Cawley et al. 1974), we made accommodations for this developmental discrepancy by approaching mathematics from a conceptual view and stressed work in geometry, measurement, patterns, sets, and fractions over rote computation and extensive work on numbers.

A second factor in programming for the mildly retarded is ultimate level of attainment. Generally speaking, when mildly retarded children leave school, they demonstrate proficiency in mathematics that is akin to that found in third and fourth grades. Figure 1.2 shows comparisons between two groups of mildly retarded children (Cawley and Goodman 1968). As can be seen, the primary area of growth is in computation. Reasoning and concept scores show only a modest increment. Special educators have tended to compensate for the limitations in ultimate levels of attainment for the mildly retarded by emphasizing applications and real-life problems. Unfortunately, many of these real-life problems were not very real. Tasks such as the completion of a long form for income tax purposes, the reconciliation of a checking account,

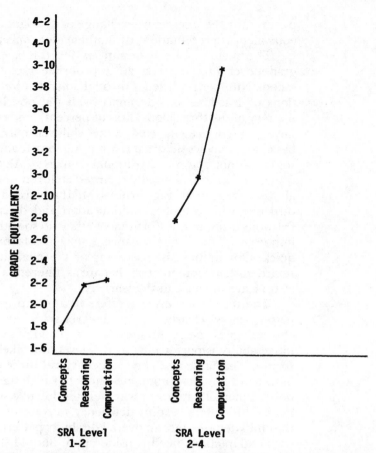

FIGURE 1.2 Mathematics achievement profiles for mildly retarded children

and interest on payments were never very realistic. This is an important consideration, given that present-day mildly retarded children are below the developmental levels of children with similar labels a decade ago.

Meyen (1968) compared mildly retarded and nonretarded children by determining the age at which 60 percent of the children in either group performed mathematics items correctly (Figure 1.3). In the main, the mildly retarded children were three to five years older than nonretarded children when 60 percent of the sample responded correctly to an item. Note the great diversity in the types of items Meyen studied and their curriculum implications.

Levine and Langress (1985) show that mildly retarded adults do generate functional problem-solving skills in everyday circumstances. These investigators studied the everyday cognition of mildly retarded adults in a supermarket. They observed their subjects using different cognitive strategies when shopping. Their most competent shoppers used elaborate schemes for estimation, calculation, and comparative shopping. Competent shoppers also used a shopping list. Only four subjects were able to use their arithmetic skills to total their bills. This is likely to be true of most regular shoppers, particularly if the list is long. Some shoppers bought the same things every time they went shopping. Others paid in larger currency to get change rather than attempt to calculate the exact amount. Ironically, one of the points made by these researchers was that the relative effectiveness of the mildly retarded in shopping may hamper the development of other problem-solving skills.

The supermarket is an interesting arena in which to observe and develop problem-solving skills. Other forums should be incorporated into the program to provide appropriate applications. Sitting in a classroom filling out

Skill or concept	Mildly retarded	Nonretarded
1. The meaning of one-third	10.7	8.0
2. Applying concept of one-half to a group	13.7	8.0
3. Relative value of fractions in sequence	—	—
4. Meaning of one-half	11.8	8.0
5. Relative size of fractions	—	10.3
6. Design of shapes	11.7	8.0
7. Term *triangle* to shape of a triangle	11.8	8.0
8. Term *square* to shape of a square	13.6	8.0
9. Distinguishing numerals from signs	12.1	8.0
10. Relative value of numbers	15.1	8.0
11. Computing addition facts	12.3	8.0
12. Distinguishing odd from even numbers	16.1	8.0
13. Placing numbers into sequence	12.6	8.0
14. Finding remainder through division	—	9.8
15. Relating addition sign to the word *add*	11.0	8.0
16. Identifying whole numbers	14.3	8.0
17. Reading Roman numerals	—	8.9
18. Rounding to nearest ten	—	8.5
19. Relating decimals to fractions of objects	—	11.4
20. Computing percent problems	—	11.2
21. Writing numbers	13.4	8.0
22. Determining value expressed in decimals	—	11.7
23. Reading numbers (602)	13.2	8.0
24. Computing averages	—	11.3
25. Counting objects	11.6	8.0
26. Computing the operations of division	—	11.6
27. Computing subtraction facts	13.5	8.0
28. Relating periods of time	—	10.5
29. Counting by 5's	12.0	8.0
30. Estimating height	15.8	8.7
31. Relative value of dollars	10.7	8.0
32. Counting money	11.9	8.0
33. Relating value of a dime to 10 pennies	12.3	8.0
34. Writing amounts	13.2	8.0
35. Computing value of coin equal to a quarter	13.4	8.0
36. Making change for a dollar	15.3	8.6
37. Making change	13.2	8.0
38. Counting change	17.3	9.8
39. Quantity buying	14.3	8.2
40. Computing price differences	15.5	8.0
41. Estimating change	—	10.2
42. Computing costs	12.7	8.0
43. Knowing months of year in sequence	11.5	8.0
44. Relating a quarter of an hour to 15 minutes	—	8.0
45. Relative value of lines	12.3	8.0
46. Reading length from a ruler	—	8.5
47. Reading time from a clock	14.9	8.4
48. Relating 1 pound to 16 ounces	16.6	8.0
49. Relating 1 dozen to 12 units	12.5	8.0
50. Determining the number of days in a week	13.3	8.0
51. Concept of 1 foot in inches	13.9	8.0
52. Computing a problem involving the subtraction of pints from quarts	—	9.4
53. Determining amount of elapsed time when two specific times are given	17.2	9.4
53. Reading time in written form	—	11.5
55. Reading thermometers	—	8.0

FIGURE 1.3   Ages at which mildly retarded and nonretarded children learn certain tasks

checks in checkbooks, balancing budgets, and the like cannot provide the same opportunities for learning and application as can be found in community-based programs.

The third factor to consider in planning for the mildly retarded is that reasonable estimations of academic growth suggest one year of achievement for every two years of school or two years of achievement for every three years of school. Curriculum planning mandates the inclusion of such developmental considerations. Tactics such as having the mildly retarded repeat a grade are clear indications of a lack of planning on the part of the school system.

The line marked *far below average* in Figure 1.1 illustrates the plight of children who are inordinately disabled in mathematics. This group could include children with high cognitive abilities, average mental abilities, or abilities at the mildly retarded level. Their progress in mathematics is extremely slow or stagnant at the earliest of levels. Fitzmaurice-Hayes (1985) describes severe learning disability in mathematics as failure to attain a level of performance consistent with ability; the important consideration is a concomitant resistance to ordinary instruction. However, ordinary instruction is what they generally receive. In effect, children who are far below average require a very special appraisal (Cawley 1985b). These are children whose needs are seldom met in traditional school programs.

Comparisons among Handicapped Children

Grise (1980) studied the performance of mildly handicapped children on Florida's minimum competency test. His data for the mathematics section of the test show:

	1977 Percent passing	N	1978 Percent passing	N	1979 Percent passing	N
ED	17	114	42	190	27	136
SM	25	79	48	33	50	32
LD	17	502	27	652	30	923
MR	1	479	9	81	3	652

Given that mildly handicapped children can be excused from the state competency test, those actually taking the test may represent a biased sample. Regardless, the performance of mildly retarded (MR) youth at the eleventh-grade level is a clear indication that they reach a ceiling in mathematics below that expected on a test of minimum competency. Some 70 to 80 percent of the learning disabled (LD) do not progress to a level established for minimum competency. Performance among the socially maladjusted (SM), called behaviorally disordered in this text, shows higher levels of attainment on the minimum competency test.

Table 1.2 was constructed from data on the performance of selected groups of mildly handicapped children on Louisiana's state competency test. The data show the percent passing at each grade level. The Louisiana data are compiled on children who are enrolled in regular classes. No children in self-contained classes are included. It is known that some members of the present sample are two to three years above age for their grade placement. Thus, the performance levels are likely to show more optimum levels of attainment for these samples than for the population of mildly handicapped children as a whole.

The pattern of responses shows that the number of children passing in mathematics at the elementary level exceeds those passing in reading. At the upper grade levels, the pattern reverses itself: the number passing in reading is greater than the number passing in mathematics.

The state of North Carolina requires that a high school competency test

TABLE 1.2 Percent of handicapped and nonhandicapped children passing Louisiana basic competency test

	80–81 Read	80–81 Math	81–82 Read	81–82 Math	82–83 Read	82–83 Math	83–84 Read	83–84 Math
Grade 2								
Mildly retarded			65	68	69	67	71	71
Emotionally disturbed			76	83	82	82	81	86
Learning disabled			80	86	83	86	84	87
Nonhandicapped			92	92	94	92	94	93
Grade 3								
Mildly retarded					52	56	65	68
Emotionally disturbed					79	82	82	82
Learning disabled					75	78	78	79
Nonhandicapped					89	87	90	88
Grade 4								
Mildly retarded							51	56
Emotionally disturbed							77	79
Learning disabled							74	78
Nonhandicapped							89	88
Grade 7								
Mildly retarded	34	28	34	38	40	31	42	32
Emotionally disturbed	56	41	61	41	63	42	65	48
Learning disabled	58	44	62	48	63	48	64	49
Nonhandicapped	77	64	80	66	82	68	83	70
Grade 10								
Mildly retarded	33	30	34	30	39	33	40	32
Emotionally disturbed	59	48	66	48	65	50	59	48
Learning disabled	58	55	60	53	60	53	60	53
Nonhandicapped	77	72	77	66	78	71	80	72

must be passed if a student is to receive a high school diploma. The test is administered in the eleventh grade and may be administered repeatedly to students who fail. Table 1.3 shows the results of the North Carolina test for 1979 (McKinney and Haskins 1980) for learning-disabled and mildly retarded students.

Developmentally, only 3 mildly retarded children with IQs below 50 passed either the reading or the math test. Of those with IQs above 75, 31 percent passed the reading test and 26 percent passed the math test. Among LD students, 92 percent of those with IQs above 95 passed reading and 68 percent passed math. Program changes were made for some students who failed the test. On a subsequent testing, 24 percent of the MRs who had program changes passed the reading test, in contrast to 11 percent who did not have changes. For LD students, about 50 percent of those with changes and those without changes subsequently passed the reading. The result for mathematics was considerably less for both groups.

The pattern of performance among secondary school handicapped children in mathematics and reading is not unlike that shown for nonhandicapped children. Table 1.4 displays data for the Connecticut ninth-grade proficiency test of October, 1980. Of the nearly 41,000 students taking the

TABLE 1.3 Percent of LD and MR children passing state competency test

	Reading N	Reading Percent	Math N	Math Percent
Mentally retarded	1890	12	1887	7
Learning disabled	652	56	652	47

TABLE 1.4 Performance of children on ninth-grade proficiency test

Percent of items correct	
Math	
Computation	78.0
Concepts	70.7
Problem solving	73.0
Total	**74.2**
Language arts	
Mechanics	81.2
Composing	82.3
Library	77.7
Total	**80.8**
Reading	79.5
Percent of children passing	
Reading	91.4
Language	90.8
Math	74.6

test, about 91 percent passed reading and about 75 percent passed mathematics. A summary of performance data from competency tests administered to the mildly handicapped in three states shows:

- Performance in mathematics is higher in elementary grades.
- Performance of all target groups is less satisfactory in mathematics than in reading at the secondary level.
- Pattern of performance is reasonably consistent from year to year.
- Mildly retarded children perform substantially below learning-disabled and behaviorally disordered in both reading and mathematics.

IMPLICATIONS

Given the consistent tendency for differences in mathematics development among the mildly handicapped, it is evident that more than one concept of programming is needed. In our own work on *Project MATH* (Cawley et al. 1974), we began with the fact that children attend school for some twelve to fourteen years, and during that time they are entitled to a reasonable representation of mathematics.

Accordingly, *Project MATH* contained curriculum strands in patterns, sets, geometry, fractions, numbers, and measurement. Both concepts and skills were stressed, with the former providing the major influence. *Project MATH* also included extensive problem solving (see Chapters 9 and 10) and opportunities for social application through forty-five units of study on real-life problems. To provide additional practice and opportunities for generalization, *Project MATH* contained about 4500 games, each coded to specific mathematics topics.

Multi-modal-mathematics (Cawley et al. 1980) differed from *Project MATH* in that each of the four major strands—geometry, numbers, measurement, and fractions—were developed independently so that specific needs of children in the upper grade levels could be given attention. Thus, if the child had an unusual difficulty in fractions, a program in fractions was available to provide curriculum and instructional options. Over 300 problem-solving guides were included in *Multi-modal-mathematics*.

Each of the above programs was guided through a variety of instructional options via the Interactive Unit (IU) (see Chapter 2). The IU provides a systematic yet flexible system of sixteen teacher-learner interactions for the teaching of mathematics. In addition, ditto masters or worksheets can be made from ten of the sixteen combinations.

A recent program, *Learner Activity Program: Developmental Mathematics* (Cawley et al. 1987), provides the teacher with single worksheets for 110 mathematics concepts and skills and shows the teacher how to develop other worksheets on the same concept through the use of the Interactive Unit. This provides teachers with a means to develop varying representations for a single concept or skill to reinforce the child or to expand generalizations. In use, a regular class teacher could review a topic with a special education teacher, and either or each could develop different worksheet formats to meet the needs of mildly handicapped children.

We strongly believe in long-term planning for curriculum and instruction. This type of planning requires that one incorporate responses to five basic questions:

1. What shall be taught?
2. How shall it be taught?
3. When shall it be taught?
4. In what sequence shall it be taught?
5. For how long shall it be taught?

Although short-term efforts certainly have their value, it is our belief that the mildly handicapped are in need of and entitled to comprehensive efforts by the professionals who serve them. We trust that the remainder of this book will be consistent with these beliefs.

LEARNING AND INSTRUCTION

INTRODUCTION

How do humans learn? Many answers to that question have been posed. We assume new behaviors by modeling the behaviors of others; we depend on reinforcement to cement new behaviors; we acquire knowledge from being told information; we learn through discovery; we actively build the cognitive structures that represent true understanding. This list represents only some of the many different positions; there are many interrelationships.

How do teachers synthesize different positions to achieve a coherent, valid set of teaching behaviors? Some adopt a single position that best suits their personal style and conviction. Others, perhaps most, elect a more eclectic approach, sampling from several schools of thought. The latter choice requires a consciousness of what theory is operative at any one time and a rationale for the combination of methods used.

Cawley (1984a), in the Interactive Unit, formulated a model through which the actions of both instructor and learner could be organized and monitored. This chapter has as its focus the Interactive Unit and its potential for giving expression to different hypotheses about the learning process. How do the component interactions reflect time-honored, research-verified explanations of how and why people learn?

THE INTERACTIVE UNIT

According to Cawley (1984b), the Interactive Unit is represented by a matrix like that illustrated in Figure 2.1.

The sixteen cells "format the interactions between teachers and learners or among teacher, learner, and material" (Cawley 1984c, p. 241). The following set of lessons illustrates the use of the Interactive Unit. The objective is:

> When shown a 30°–60° right triangle, and given the measure of the short leg of the triangle, the student will supply the measure of the hypotenuse of the triangle.

The materials needed are pieces of string or strips of paper, several representations of 30°–60° right triangles, rulers, paper, pencil, and worksheets as described in the sixteen activities that follow.

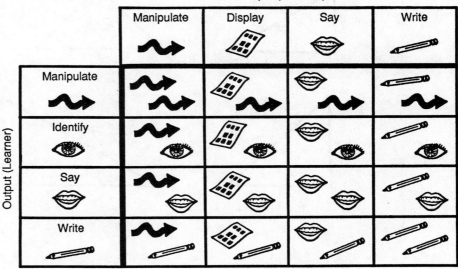

FIGURE 2.1 The Interactive Unit

ACTIVITY 1

MANIPULATE/MANIPULATE

The instructor places several representations of 30°–60° right triangles in front of the learner. The instructor then uses string or a strip of paper to measure the hypotenuse of one of the triangles, cutting the string or strip of paper so that its length corresponds to that of the hypotenuse. The instructor demonstrates that the length of the same piece of string or strip of paper folded in half matches the length of the short side of the triangle. The instructor asks the student to do the same with another of the triangles. If the student responds correctly, the instructor presents another triangle for the same treatment. If the student does not respond correctly, the instructor models the behavior again and repeats the request for imitation.

ACTIVITY 2

MANIPULATE/IDENTIFY

The instructor repeats the input to the first activity. The student selects from representations of various kinds of triangles those that seem to demonstrate the relationship the instructor is illustrating.

ACTIVITY 3

MANIPULATE/SAY

The instructor repeats the input to the first activity. The student is asked to describe what the instructor is illustrating. The desired response is a statement to the effect that the hypotenuses (long sides) of the triangles under consideration are twice as long as the short legs. Any response similar to that should be encouraged.

ACTIVITY 4

MANIPULATE/WRITE

The instructor gives the learner a copy of a worksheet such as that illustrated in Figure 2.2.

The instructor repeats the activity described previously. The learner is required to point to or mark all the triangles that illustrate the property that the instructor is illustrating.

Learning and Instruction

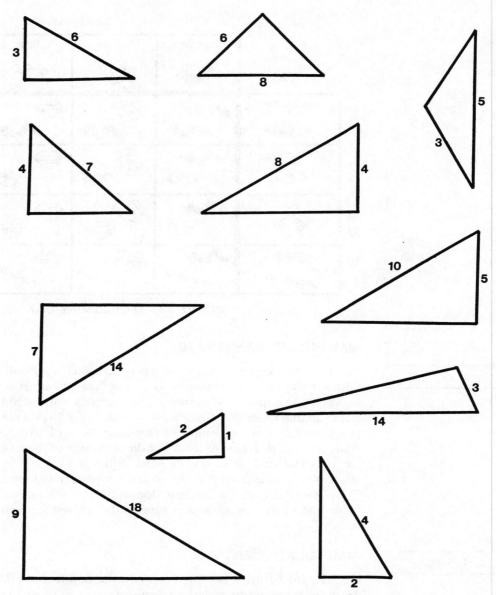

FIGURE 2.2   Worksheet: 30°–60° right triangles

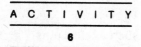

DISPLAY/MANIPULATE

The instructor places several representations of 30°–60° right triangles in front of the learner. On each triangle the hypotenuse and the short leg are highlighted. The learner is requested to use string or strips of paper to illustrate the relationship between the longest side and the shortest side of each triangle.

DISPLAY/IDENTIFY

The instructor places representations of two line segments, one twice as long as the other, before the child. The learner is given a worksheet on which are represented several samples of 30°–60° right triangles. (See Figure 2.3.) The learner is asked to point to or mark the two sides in each triangle that show the relationship represented by the two line segments.

FIGURE 2.3   Worksheet: Display/identify

A C T I V I T Y

7

DISPLAY/SAY

The instructor places a representation of a 30°–60° right triangle like that used in the preceding activity before the learner. The learner is asked to describe the relationship between the two sides that are highlighted.

A C T I V I T Y

8

DISPLAY/WRITE

The instructor places several representations of 30°–60° right triangles before the learner, or presents the learner with a worksheet; for example, one like that represented in Figure 2.3. The learner is requested to use a ruler to measure the short leg and the longest side of each triangle, and to write the measures close to the lines representing the respective sides.

A C T I V I T Y

9

SAY/MANIPULATE

The instructor places several representations of 30°–60° right triangles in front of the learner and tells the learner that the measures of the two smaller angles are 30° and 60°, while pointing to each angle. The instructor points

FIGURE 2.4   Worksheet: More 30°–60° right triangles

out that in such a triangle, the measure of the hypotenuse (the longest side) is exactly twice that of the shortest leg. The instructor provides the learner with string or strips of paper and asks the learner to demonstrate that relationship.

SAY/IDENTIFY

The instructor explains to the learner that in a 30°–60° right triangle the longest side (hypotenuse) has a measure twice that of the shortest leg. The learner is given a worksheet on which many types of triangles are represented, among them several examples of 30°–60° right triangles (see Figure 2.4). As the instructor dictates measures of sides, the learner is asked to mark all those triangles in which the longest side has a measure that is twice as long as the shortest side.

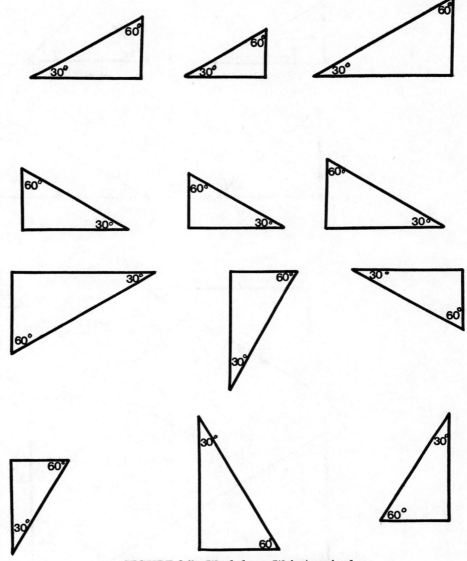

FIGURE 2.5   Worksheet: Write/manipulate

SAY/SAY

The instructor explains to the learner that in a 30°–60° right triangle the longest side (hypotenuse) has a measure twice that of the shortest leg. The instructor asks the learner to repeat the relationship. The instructor then gives examples such as: "Suppose in a 30°–60° right triangle the shortest side has a measure of 12 cm. How long will the longest side be?" "Suppose in a 30°–60° right triangle the shortest side has a measure of 3 inches. How long will the hypotenuse (longest side) be?"

SAY/WRITE

The instructor provides the learner with paper and pencil. Explaining that all of the triangles under consideration will be 30°–60° right triangles, the instructor tells the learner that measures of the shortest side of such a triangle will be dictated. The learner is to write down the measure of the longest side

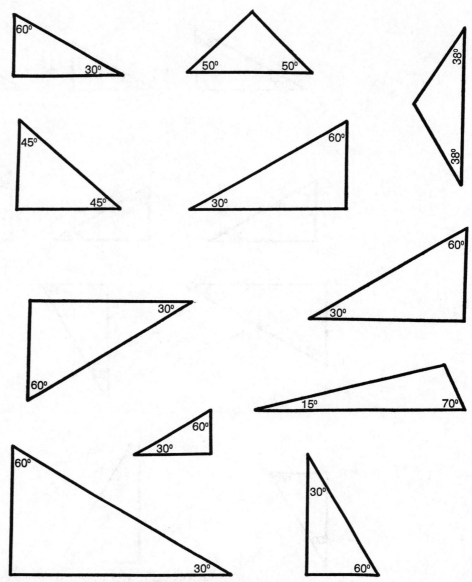

FIGURE 2.6   Worksheet: Write/identify

(hypotenuse). The instructor dictates measures such as 3 inches, 5 centimeters and so on.

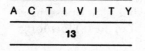

WRITE/MANIPULATE

The instructor provides the learner with an activity sheet like the one represented in Figure 2.5. Using string or strips of paper, the learner is to demonstrate that in each triangle represented, the hypotenuse has a measure which is twice that of the short leg of the triangle.

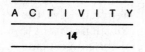

WRITE/IDENTIFY

The instructor provides the learner with an activity sheet like the one represented in Figure 2.6. The learner is to mark all those triangles in which the longest side has a measure which is twice that of the shortest side.

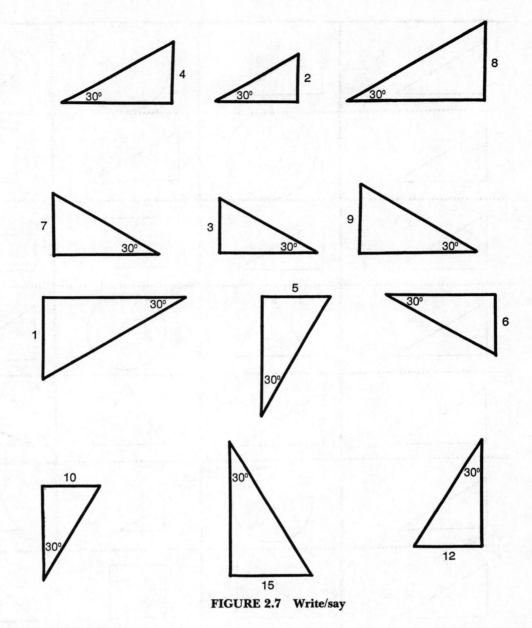

FIGURE 2.7   Write/say

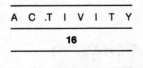

WRITE/SAY

The instructor provides the learner with an activity sheet like the one represented in Figure 2.7. The learner is to state the measure of the hypotenuse of each right triangle.

WRITE/WRITE

The instructor provides the learner with an activity sheet like the one in the previous activity. The learner is to write the measure of the hypotenuse of each triangle.

As described above, the sixteen activities illustrate the application of the Interactive Unit to the content selected. We now examine the ways in which different combinations can reflect different approaches to the tasks of instruction and learning.

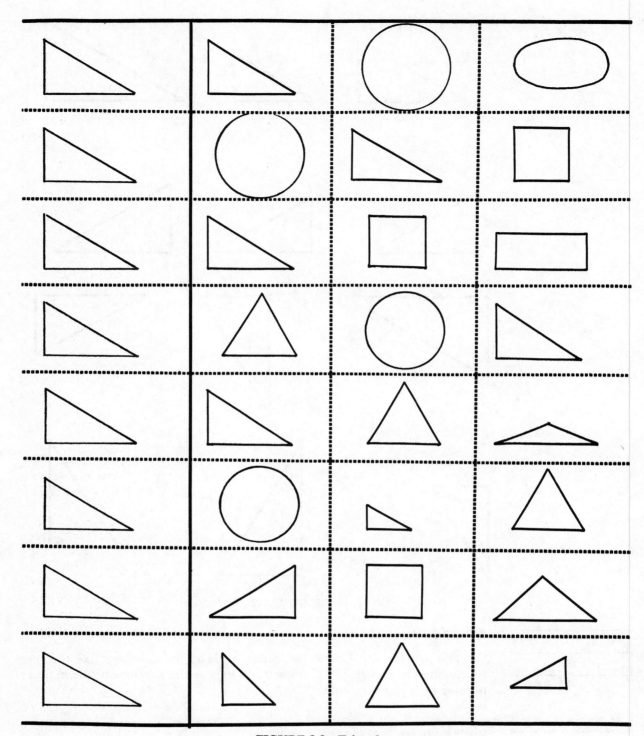

FIGURE 2.8 Triangles

LEARNING AS CHANGES IN OBSERVABLE BEHAVIOR

A behavioral approach to the teaching task generally reflects several assumptions: teaching primarily requires the setting up of conditions that will lead to desired behaviors that can be reinforced, followed by the reinforcement of those behaviors; simpler types of learning precede more sophisticated kinds; complex tasks consist of a series of discrete subtasks; the learner takes a passive role, either modeling the behavior of others or responding to reinforcement.

The sixteen activities outlined previously, combined or sequenced in different ways, provide for all those assumptions. The manipulate/manipulate, say/say, and write/write combinations can all be used to elicit desired behaviors and provide amply for the demonstration of behaviors that the mildly handicapped child is to model. Within each interaction there can be designed lessons that elicit desired behaviors and take the child step by step through the requirements of a behavior.

Examine the set of display/identify tasks illustrated in Figure 2.8. Each task might represent an item taken from a worksheet consisting of items of similar difficulty.

The content of the activities, the 30°–60° right triangle, is a prerequisite for the achievement of the objective around which the previous sixteen activities were designed. The first item almost forces the student to select the right choice, thereby providing the instructor with an opportunity to reinforce the response. With each item the required discrimination is greater, either because of the number of choices, the similarity among the choices, or the orientation of the correct response. The last task requires the child to select the correct response from among triangles, two of which are right triangles. The desired response is the mirror image of the model, and the desired response has a different size from that of the model. Within this one combination, the child has progressed from a fairly gross level of discrimination to a sophisticated one.

LEARNING BY LECTURE

The telling or lecture approach is regarded by many as less than desirable for effecting true learning. It is nevertheless an efficient instructional method if care is taken to ensure comprehension of the content being imparted. The Interactive Unit provides the say or write input, which allows the imparting of information through speaking or writing. The four output behaviors can be used to determine the level of understanding, as the following set of activities illustrates.

Say input.   The instructor provides orally an explanation like the following. "The word *twice* means the same as two times as much as. This red stone weighs twice as much as that blue stone means the weight of the red stone is two times that of the blue stone. I am twice as old as you are means my age is two times your age."

Manipulate output.   The instructor provides the student with some colored beads or strips of paper of different lengths. The instructor gives directions like these: "Make a group of blue beads. Now make a group of twice as many red beads. Find a short piece of paper. Now find a piece of paper twice as long as the first one."

Identify output.   The teacher provides the child with a worksheet like that illustrated in Figure 2.9. In each row the child is to mark the response that is twice the size of the figure to the left of the row.

Say output.   The teacher poses questions like these: What number is twice three? If a blue book weighs twice as much as a brown book, and the brown book weighs six ounces, how much does the blue book weigh? If a line

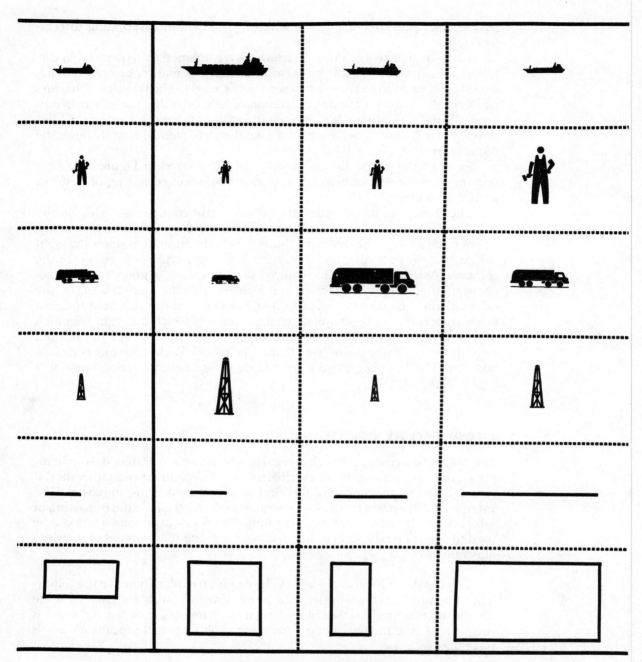

FIGURE 2.9 "Twice as long"

segment is five centimeters long, how long will a line segment of twice that length be? The student is to answer each question orally.

Write output. The teacher dictates a list of numbers. For each number the student writes the number that is twice the given number.

The child who successfully performs the tasks described for each of the four outputs demonstrates an understanding of the term *twice as*. Since there are many times when we do receive information orally or in writing, mildly handicapped children need to learn how to check themselves on their understanding of such information. Activities like those described above, then, can also perform a "learning how to learn" function.

LEARNING AS AN INTERNAL PROCESS

Until now we have looked at learning as equivalent to changes in observable behavior. For many educators, such a description of learning falls short of the real phenomenon. For these theorists, learning is an internal process, described in different ways. Accordingly, the role of instruction takes on different emphases.

Bruner (1963) described three systems of skills that the mature learner seems to have developed. He labeled these systems as enactive, iconic, and symbolic. These skills are used by the learner to create internal models of the external environment. The skills are usually developed in the order listed above, but all three are intact in the adult learner.

The enactive mode of representation is manipulative or motoric in nature. The child learns through manipulation, without the use of pictures or words. To get some idea of the meaning of this enactive mode, think of riding a bicycle or roller skating. Generally speaking, one learns either skill principally by doing; neither a picture nor verbal explanation can substiture for the sense of balance needed for each activity.

Iconic representation implies the use of pictures or some other form of internal imagery. Upon hearing the word *triangle,* one generally conjures up an image of a shape, not a definition in words. Adults often know many things iconically—the face of someone dear, the appearance of one's car, the "meaning" of the botanical term *rose,* the experience of being in a redwood forest. There are many instances in which words serve only to cloud an issue; on the other hand, this "iconic" knowledge is often so image-bound and altered by personal experience that in a particular case, people disagree on the appearance of something all have seen. Witnesses to an accident often fall into this group.

The third type of representation—symbolic learning—depends on the use of language as a medium of thought. Language provides the capability to deal with the hypothetical—or what might or what might not exist—as well as what does. As an underpinning for memory, words serve as a "permanenting" agent for experience; language also provides the principal tool of reflective thinking.

The three modes of representation are hierarchical in two senses. They appear in a person's repertoire in the order given previously. Very young children engage primarily in enactive learning. Between the ages of five and seven, reliance on iconic modes of representation seems strongest. With the approach of adolescence, language becomes more and more important as the vehicle for thought. The sequence is also one of increased knowledge—that is, one is better off being able to employ the symbolic mode of representation than being confined to the enactive or iconic modes.

The implications of these three modes of representation for instruction seem clear. All three types of learning must be taken into account during any instructional endeavor. The function of teachers is twofold: they must make decisions about which modes are appropriate for a given set of content; they must provide appropriate transitions from one mode of representation to another where such transitions are desirable. Not all experiences need to be represented symbolically. The runner waiting at the starting line for the signal to be off does not need to be reciting to himself or herself the physiological changes involved in the act of running, although the physician on the sidelines might well be able and inclined to discuss those changes.

The reader can readily perceive the potential of the Interactive Unit for developing instructional sequences that provide for the three modes of representation. Bruner (1963) was concerned with the learner's internal representations; the Interactive Unit also regulates the instructional input on which

FIGURE 2.10   Discovering a relationship

the learner is to build his or her own representations. Hence input activities can be categorized not only as visual, tactile, or oral in nature, but also as iconic or symbolic. The learner's output, seen as giving evidence of internal representation, can be classified as enactive, iconic, or symbolic.

Akin to the preceding, but worthy of separate discussion, is the type of learning sometimes called *discovery* learning. While not confined to inductive reasoning, discovery learning often finds its expression in the child's arriving at a general conclusion after the examination of several specific instances of the fact. The Interactive Unit supports such an instructional approach quite easily, as the following set of activities demonstrates.

A C T I V I T Y

1

DISPLAY/WRITE

The instructor gives the student a worksheet on which are drawn several models of 30°–60° right triangles like those in Figure 2.10, and a ruler and a

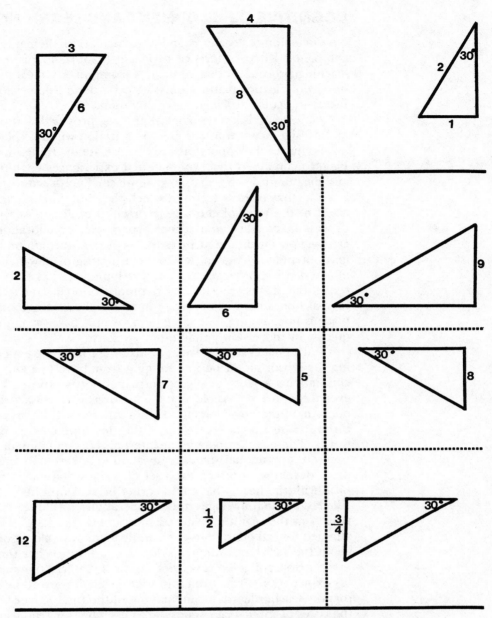

FIGURE 2.11   Predicting measures

protractor. The student is to measure the acute angles in each triangle, the hypotenuse of each triangle, and the short leg of each triangle. Each measure is to be recorded. When a student notices a pattern, he or she is to write a sentence about the pattern.

WRITE/SAY

The student is given a copy of a worksheet like that shown in Figure 2.11. After examining the triangles at the top of the page, the child is encouraged to predict the measure of the hypotenuse of the triangle in the first frame. If the guess is correct, the student continues to state the measures of the hypotenuses of the remaining triangles. Finally, the student is asked to formulate and state the rule he or she has been using.

COGNITIVE DEVELOPMENT AND LEARNING

No discussion on learning, and in particular, mathematics learning, would be complete without reference to Piaget (1959) and his interpretation of the learning process. In that context, learning in any meaningful sense depends on several factors: maturation, physical experience, socialization, and a fourth factor—peculiar to Piaget—equilibration.

Maturation is a physiological process; provided that proper nourishment and no significant traumas exist, maturation will take place. Physical experience requires the opportunity to manipulate the things in one's environment. Piaget maintained that without such experiences cognitive development in any meaningful sense cannot take place. The process of socialization is defined as that process through which the young people in a society are educated to the ways of the society or culture. For most western cultures this process takes place as a school experience. Equilibration is the process of creating new mental structures that represent the attainment of new learning and potential for moving on to the acquiring of new structures.

According to Piaget, mental development takes place in a series of four sequential stages: sensorimotor, preoperational, concrete operational, and formal operational. The order in which these stages of development take place is unchanging, but some persons may pass through some stages more quickly or more slowly than other persons.

During each stage, certain kinds of learning are expected to occur. During the sensorimotor period, ranging from birth to about the age of two, the child attains a sense of object permanence (the knowledge that things exist even if they are not visible) and the motor skills necessary to function in and to operate on the environment. The preoperational years, from about age two through age six or seven, mark the development of internal images and words. These make possible the internal representation of reality. During the concrete operational stage, from about age seven to age eleven or twelve, there develops an internalized set of actions that allows the child to accomplish with the head what could earlier be accomplished only with the hands. The last of the four stages, formal operations, is characterized by the ability to reason in the manner required by scientific thinking. The person who has attained formal operations can handle many variables at once and can reason about the variables without having to check his or her thinking.

The foregoing is a very brief sketch of material that has required volumes for explanation. From even this brief sketch, however, the reader will observe that the teacher/learner combinations of the Interactive Unit provide for both the types of input Piaget considered essential for true learning to take place and for children in different stages of cognitive development.

AFFECTIVE CONSIDERATIONS

Successful mathematics programming for the mildly handicapped child demands not only constant awareness of the teaching/learning theories operative at any one time but also routes of access that minimize the obstacles to such a child's learning. The Interactive Unit provides those means of access. The child who can neither read nor write can acquire mathematics concepts through tasks that do not require reading or writing. There are nine such combinations: manipulate/manipulate, identify, say; display/manipulate, identify, say; and say/manipulate, identify, say. Postponement of mathematics instruction is not necessary.

The teacher dealing with a student who neither speaks nor understands spoken English to any great degree can use all combinations except the manipulate/say, display/say, say/manipulate, identify, say, write, and write/say. There remain nine other combinations through which mathematics content can be conveyed. Because mathematics symbols are more universal in nature than words, the write input can be used as a vehicle for instruction, as long as directions are provided through pictures or mathematical symbols. Once again, mathematics instruction need not be delayed until the student attains a facility with English.

Affective concerns assume a large role in the teaching/learning process. The Interactive Unit also affords opportunities for instructional management that is ordered to certain affective goals. By their very nature, certain combinations of instructor/learner interaction promote different forms of behavior. Manipulation input, for example, requires the physical proximity of instructor to learner. Such proximity lends itself to the quiet word of praise, the guiding of motor activity, and, most importantly, to the masking of failure.

To understand the potential for masking failure more clearly, imagine that you, as an instructor, are showing a child how to use a ruler to measure line segments. After your demonstration of proper placement of the ruler, the child fails to place the ruler correctly to measure the line segment. With little effort or fuss, you can rearrange the materials being used so that the desired behavior is ensured. Compare this situation with a workbook page on which a set of line segments are drawn. The child who writes an incorrect measure on the first item is faced with the mistake for the rest of the page. The manipulate input helps to avoid this penalty until the child is likely to make the correct response. Failure is easily masked and successful experiences substituted.

Other input/output combinations serve other purposes. The mildly handicapped child for whom group work is detrimental, either to himself or herself or to the group, can participate in several other combinations. These include display input combined with identify output or write output; write input combined with identify output or write output; and even say input, identify, or write output. The latter type of activities can be used to provide experiences through which the child is gradually brought into working relationships with other students.

For some mildly handicapped children, the say output might be a source of embarrassment and humiliation if they do not know the answer. For the children, the opportunity to express an answer by way of some other output may result in greater self-confidence and the courage to take a stab at an answer aloud in the presence of other students. These examples summarize but a few of the affective benefits that can be derived from utilizing a model like the Interactive Unit for instruction.

ALTERNATIVE REPRESENTATIONS

The Interactive Unit provides a means for presenting children with alternative representations of mathematical concepts and skills. Historically, the field of special education has viewed multimodal instruction from a hierarchical (or preferred modality) perspective. That is, some children were thought to be visual learners, and others were thought to be auditory learners. It was thought that if one could identify auditory or visual learning strengths or weaknesses, one could select instructional procedures that could access the learner through strengths or design instructional procedures that would remediate the weaknesses. Whether or not the preferred-modality orientation will be fully validated is not a primary consideration from our perspective. We do not view the use of alternative representations as hierarchical. Of greater importance to special education is the recognition that the Interactive Unit

Teacher	Learner
1. Place the 2 sets of 3 objects. Say, "Do as I do." Repeat with 3 sets of 4 objects. Repeat with 5 sets of 2 objects. As you are doing each of these, say, "I have _____ in each set. I have _____ sets. I have _____ sets with _____ in each set."	Does what teacher does Says what teacher says

2. Put out 2 rows with 4 in
each row. z z z z
           z z z z
Say, "I have two rows.
     I have four in each row.
     I have two (point) times four."
Have child say what you say.
Repeat with:
    2 rows, 6 to a row
    5 rows, 4 to a row
    2 rows, 10 to a row
    5 rows, 7 to a row
    2 rows, 8 to a row
	Does what teacher does
	z z z z
	z z z z

3. Say, "Listen, make two rows with three in each row."
Repeat with:
    2 rows, 7 to a row
    5 rows, 3 to a row
    10 rows, 4 to a row
    10 rows, 6 to a row
    6 rows, 10 to a row
	Makes rows as stated

4. Say, "Watch me."
Make   z z z
           z z z
Now this z z z
           z z z
           z z z
           z z z
           z z z
Point to a 2×3 and ask,
"What does this show?"
 	
Point to 5×3 and ask,
"What does this show?"
Say, "Watch me," and push
2×3 and 5×3 together.
Tell child to do same.
"See this." (Point to new
arrangement.) "Can you say it
for me?"
 	
"OK, what's happened? If
I put 2×3 and 5×3 together,
I get 7×3. You say it for me."
Repeat with:
    2×4 plus 5×4
    2×8 plus 3×8
    5×7 plus 1×7
    10×3 plus 2×3
	Makes one like teacher
	z z z
	z z z
	z z z
	z z z
	z z z
	z z z
	z z z
	
	"Two rows, three to a row" or
	"Two times three."
	
	States response
	
	Makes one like teacher
	
	Expected response is
	"7 rows of 3 or 7×3"
	
	States response

5. "Good. Now watch me."
Do this: z z z z
           z z z z
           z z z z	

Figure 2.12 continues

Teacher	Learner

z z z z
z z z z
"Make one like mine."
"What does it show?"
"Now watch me."
Separate the 5 × 4 into
  2 × 4   z z z z
             z z z z
and 3 × 4   z z z z
             z z z z
             z z z z
	Makes array as shown
	"Five times four"

"You do it." "Tell me, what did I do?"	Makes arrays as shown "You had 5 × 4 and you changed it to 2 × 4 and 3 × 4."

Repeat with:
  10 × 3 to 5 × 3 and 5 × 3
  8 × 4 to 6 × 4 and 2 × 4
  5 × 3 to 3 × 3 and 2 × 3
	States response

"What have we? If I have this (point to 5 × 3), and I take this away (remove 3 × 3), what is left?"
Repeat, stressing whole/part relationships. Use:
  5 × 6 to 3 × 6 and 2 × 6
  10 × 3 to 8 × 3 and 2 × 3
	States "2 × 3"

6. "OK you be the teacher. Make a few for me, and I will do them like you did."
Have the child do 3 or 4.
If hesitant, take through sequence and show how to be the teacher.
Do 3 or 4, joining after the separation. See that child can do both.
  8 × 3 to 5 × 3 and 3 × 3
and
  2 × 4 and 3 × 4 to 5 × 4
	Child makes/teacher does
	
	Child responds as in steps 4 and 5

FIGURE 2.12   **Distributive property illustration**

is able to guide the instructional process in the following ways.

Mathematics concepts and skills can be represented through different interactions. By doing so we are able to reduce the impact of one disability or limitation on another. As noted previously, if a child is not able to read, the interactions that utilize the write input and write output can be bypassed in the instructional experience.

The use of alternative representations provides greater opportunities to effect transfer and generalization. The sixteen different formats can be used to represent mathematics concepts and skills and to conduct problem-solving activities. Children are provided with a variety of ways to develop conceptualizations, to transfer these to new situations and materials, and to generalize.

The use of varying combinations within the Interactive Unit activate the learner, requiring active, rather than passive, participation in the learning experience. A child who is moving blocks, walking up and down a number line, and pushing various combinations of sets about to represent multiplication is involved in representing mathematical concepts, building meaning and understanding. The activities illustrated in Figure 2.12 can be used to intro-

Pictorial	Written	Demonstration	Aural
1. Signal retained	1. Signal retained	1. Signal fades	1. Signal fades
2. Fixed signal	2. Fixed signal	2. Flexible signal	2. Flexible signal
3. Rapid reception	3. Rapid reception	3. Slower reception	3. Slower reception
4. High material need	4. Low material need	4. High material need	4. Low material need
5. Nondominant school activity	5. Dominant school activity	5. Nondominant school activity	5. Dominant school activity

FIGURE 2.13   Trade-offs in modifying instructional presentations

duce the child to the distributive property of multiplication. These experiences begin to show the child some of the interrelationships among multiplication expressions and their meanings.

Mildly handicapped children have repeatedly been shown to have difficulties with short-term memory. In spite of this, the primary instructional strategy for teaching an operation such as multiplication involves rote drill and practice—a commitment to memory of the basic fact before the children have any substantive understanding of the meaning of the operation or the interrelationships among the facts. Figure 2.13 lists some of the strengths and weaknesses of the different input alternatives of the Interactive Unit.

Note that both manipulate and say have two things in common. First, the message is transmitted sequentially in each instance. The words and the actions come to the learner in a fixed sequence. Second, as soon as each element in the sequence has been presented, it fades; the learner can retrieve it only from memory. Thus, both spoken language and manipulative actions are memory-dependent.

By contrast, the messages presented to the learner by fixed displays of object or pictures or by printed symbols remain in fixed form before the learner. The learner can scan them, study them with great care and at his or her own pace, or reexamine them as desired. Whereas printed words and mathematical expressions are presented in sequence, fixed displays of pictures or objects can be viewed from many angles and often can be examined in their entirety at a single glance.

A major limitation of the fixed visual presentation through displays or written symbols is that they are not easily modifiable for the learner. That is, it is hard to change the page of a book. If the learner cannot read the material or comprehend the meaning in a diagram or illustration, it is difficult for the teacher to modify the material. By contrast, messages presented manipulatively or through spoken language are easily changed. It is essential that the advantages and disadvantages of various combinations of input and output be understood.

The interactive approach attains its highest level of effectiveness when the learner clearly understands that various means are being used during instruction and learning. In the metacognitive sense, it is important that the teacher explain the different interactions to the youngsters, taking care that they are aware of the need to monitor themselves, to ask questions and to know the requirements of performance under different interactive combinations. Many mildly handicapped youngsters fail to derive the maximum benefit from manipulative experiences because they fail to understand that these experiences are alternative representations of other information and concepts. It is not enough to instruct the child through alternative representations; the use of alternative representations must be explained to the child, and the child must understand their purpose and function.

LEARNING AND ACHIEVEMENT

Learning and *achievement* are terms that are used interchangeably in descriptions of learner status, placement decisions, and instructional planning. Operationally, the terms describe two distinct entities, and each plays a different role in planning for the mildly handicapped child.

Achievement is a representation of the status of a learner at a given time. Learning can be said to be an interpretation of the effort or strategies needed to help a child to move from one point or level to another.

The computation test given on Friday is an example of an achievement test. The child is tested at a point in time, Friday; the test is scored and a record made of the child's performance. Assume one child got 100 percent of the items correct, another 50 percent, and a third child fails all the items. We have a great deal of information about their achievement status, but what do we know about their learning?

Assume, for example, that for each of the preceding four days of the week there was a practice assignment, in which the first child got all the items correct each day. Assume further that the second child missed all the items the first day, got half the practice items correct the next two days, and got 80 percent of the practice items correct on the fourth day. Assume also that the third child was absent the first two days, then got 50 percent of the practice items right on the third and fourth days, but did not get the same items right each time.

Clearly, the learning opportunities—that is, the opportunity for each child to acquire mastery of the practice items—were not the same. In fact, knowing what we do about the history of the children during the week, it seems totally inappropriate to give each child the same test to measure achievement. The proper time to give each child the test would be after each has had the same opportunity at mastery. In effect, achievement should be measured after each child has had four consecutive practice sessions in which all items are completed correctly.

We might assume further that the practice items were all administered via worksheets and that all youngsters practiced on the same worksheets. Let us assume also that each worksheet consisted of twenty-four items, for a total of ninety-six practice items for the week. But the assumption that each child requires the same number of repetitions lacks validity in terms of learning. If learning is measured by the number of repetitions or trials or by the amount of time it takes a child to become proficient, the number of repetitions or number of items given to each child should be different.

Examine the case of the second child. On the first day this child got all the practice items wrong. Not only did this child not have the opportunity to practice the items correctly, but the response to every items was reinforced as the child got it wrong. Our third child missed two practice sessions and then got only half the items correct on the next two days. Further, this child failed to get the same item correct in more than one practice session. Any effort to focus on learning must begin with an appraisal of the number of repetitions or trials or the amount of time it takes a child to attain mastery. Such a determination can be made under three conditions (Cawley 1985b):

- Given any combination of n items, how many times must each be presented until the child responds correctly to each item?
- Given twenty minutes to work on n items, how many items can the child master?
- Given that the individual will be given n repetitions or trials to practice the items, how many items can be given to the child?

In the first condition, the baseline is the number of trials or repetitions needed to master *n* items. If, for example, this involves eight items, how many times must the learner practice these to know them?

In the second condition, time is the limitation. Accordingly, the number of items and number of repetitions that can be covered in that amount of time must be determined.

The third condition stipulates that only four repetitions will be given. One needs to determine how many items can be learned in four repetitions.

Special education planning for the mildly handicapped child is an achievement-based system. The discussions underlying such planning, however, tend to utilize the term *learning*. Yet planning does not take place from a learning-oriented perspective. The clearest evidence for this is in the use of worksheets or pages in books that provide each youngster with the same fixed frequency of practice items. We propose the use of interactive instruction as a means of diversifying instruction and providing different numbers and types of repetitions for mildly handicapped children. The acquisition and commitment to memory of facts and rules in mathematics by the mildly handicapped should be preceded by the development of concepts and the demonstration of a clear understanding of these concepts through the systematic use of alternative representations.

SUMMARY

For instruction to effect learning, certain principles of learning must be recognized in the instructional process. In this chapter we have examined some of those principles and the potential of a model such as the Interactive Unit for providing structure for the application of the principles. Mildly handicapped children present some challenges to the accomplishment of successful teaching. Those challenges serve to highlight the need for consciousness of how we are using the child's capability for learning and the need for great care in maximizing the potential for learning that exists. A model such as the Interactive Unit, used consistently, provides for both goals.

CHAPTER THREE | SPACE, RELATIONS, AND FIGURES

INTRODUCTION

The child reaches to touch, to grasp. He or she sees people and objects in certain places or locations. The child feels, bumps into, and learns about all of these "things." Among the early questions asked by parents is, "Where is Mommy?" or "Where is Daddy?" Names are attached and questions are asked; these aid in the initial learning process. Gradually the child begins to identify with real-world objects. (People are included with these objects.) He or she sees each object as having certain overall characteristics and occupying a certain place at a given time. From this information we gain our first clues in the *what, how,* and *when* of geometry. Other questions need to be asked. How does the child identify his or her "space"? What is real to the child? We can say, with a reasonable degree of assurance, that the world of the child involves three-dimensional space, objects within this space, relationships among these objects, and times when certain situations occur. It is at this point that we often begin to make our first errors in teaching the child about his or her world.

We impose a rigid structure as we identify specific geometric figures and relationships. We draw squares and ask about sides and vertices (corners). These shapes and relationships are far removed from the world of the child and should therefore not be the first items in the content of geometry to be taught to the youngsters. The teaching-learning process must be more natural, helping relate to what the learner sees, hears, and feels (touches).

Learners can see objects with relatively similar shapes, such as people. Objects of different sizes are available to view—large buildings and small buildings. Positions of objects can be described: the moon appears in the sky, or the basement is under the first floor of the house. Man-made objects and objects from nature provide appropriate examples of the elements of geometry. Talking about real objects and their overall geometric characteristics gives more clues to help in answering the *what* and *how* questions. We should not deal with specific details in geometry in the initial development, because a learner sees all of an object first, although the makeup of this totality could be part of another whole. For example, the face of a parent, although seen as a whole by a baby, forms parts of other wholes—the parent's body and the parent's person, for example. Only later does the child learn about the details. When an individual is introduced for the first time, specifics concerning this individual may escape the first-time observer or even the frequent observer.

What color are the eyes of your best friend? On what street is your school located?

In addition to teaching a child about his or her world, the geometry of shape, size, position, and relationships represents the beauty of mathematics. An understanding of why certain shapes and sizes please us is a part of geometry. Geometry is in art—a painting, a sculpture, a building—and art is geometry. Geometry is a part of our lives, and we are a part of geometry as we move in our three-dimensional space.

Goals

Drawing from the information above, we can define the major goals of teaching geometry to youngsters as follows:

1. To develop in the learner an awareness of the world around him or her and thus to provide some structure to the student's perception of the physical world.
2. To provide the learner with opportunities for reinforcing what already exists for him or her—building upon success; for analyzing, looking for specifics after initial development (an element of deductive reasoning); and for building relationships.

Instruction based on these goals can provide meaningful experiences for the learners in addition to being a welcome relief from working with numbers. This has special implications for teachers of youngsters with handicaps to learning and achievement. For such learners the content (the *what*) of geometric instruction and the appropriate methodology (the *how*) can provide very important benefits. In considering the characteristics of handicapped learners we noted that some of these students are not ready to deal with abstract symbols (letters, numerals, signs of relationships) until relatively late in their academic program. Others seem to exhibit strengths in working with spatial relationships, although deficits in arithmetic may be quite apparent. We have described a model for instruction that provided sixteen instructor-learner interactions (the Interactive Unit). Many of those interactions were characterized by a nonsymbolic approach to content. The *what* of geometry lends itself beautifully to the methodology of the Interactive Unit. Shapes, positions, and relationships are well-suited to manipulation.

Traditionally, the *why* of instruction in geometry, at least as promulgated by mathematicians and mathematics educators, has appeared to be far removed from the world and needs of most learning-handicapped students. For these students instruction in "logical" thinking has often been considered an exercise in futility. (It often has been, even in the case of "normal" or non-learning-handicapped youngsters). When we view geometry as a body of content, rather than as a means to an end (that is, when we see only with difficulty that it has any practicality at all), we find that the implications of this content for everyday living are numerous and significant. Tying a shoe, setting the table, replacing a worn washer in a faucet—all of these actions reflect the content of geometry.

GENERAL CHARACTERISTICS OF FAMILIAR OBJECTS (A THREE-SPACE VIEW)

Initial Development

To work with geometry a certain vocabulary must be developed and extended. After the child is able to state a correct name for a given object, *where* the objects are located becomes very important. One of the first directions given to a youngster is, "Come *here!*" The *here* means a certain location at a given time. This can change from one time to another. It is where a given person is located when the request is made. The person has a *relative position*.

At this point the child begins to function in three-space. The child hears and learns: "Put your toys *in* the box." "Leave your plate *on* the table." "Put your shoes *under* the bed." "Put your arm *through* the sleeve." "Go *up* the stairs." Verbal directions are given and specific actions follow. The directional words have meaning: the vocabulary of the learner is expanded. The directional words in three-dimensional space make up the beginning of the vocabulary of geometry for a learner. Note that objects are involved in each case and that the direction is in relation to the objects.

The first task in geometry for a teacher of early elementary school children is to extend the vocabulary of the youngsters by providing experiences that involve specific words and actions. When stating a fact or a direction or asking a question (verbal input), the teacher may accompany this verbal presentation with an object, a picture, or a specific body motion. The teacher demonstrates for the learners. The learner may be asked to copy what the teacher does, to identify something, or to tell what is happening. Whatever the situation may be, the learner must be directly involved.

It should be noted at this point, before specific examples are given to illustrate a task, that a teacher needs to be concerned with the future learning of the child. Instructional techniques should help provide for future learning experiences by foreshadowing them. This technique is used to get the learners working and thinking about a definition, procedure, or generalization before it is formally presented. We might talk about a box and describe its characteristics. Later this box will become a cube and possess the characteristics described previously with the box. A teacher should become aware of this technique and use it whenever possible. Each of the following examples possesses foreshadowing potential.

The examples given previously involve a direction word, a direction in reference to something: *in the box, on the table, through the sleeve.* A natural extension involves both of these ideas of direction and object to obtain a *relative position.* Learners can be asked to stand in front of another person or behind someone. A teacher should demonstrate by verbalizing and performing the appropriate action. Learners should be selected to perform the same activity. In the activities that follow, "I" stands for *instructor* and "L" for *learners.* We assume here that all students face in the same direction. To be *in front of* someone therefore means that someone sees you (or your back, to be more precise), but you cannot see him or her without turning around. The reverse is true if you are behind the person.

A C T I V I T Y

SAMPLE INSTRUCTIONAL ACTIVITY

I: Ask at least three people to get in a line, facing one side of the room, with the second person behind the first person and the third person behind the second person. (Let's say that Mary is the first person, Joe the second person, and Beth the third.)

Tell the rest of the class that you are going to ask questions about where Mary, Joe, and Beth are standing. (State whether you want the answers to be given by all the class or by one individual. Note which learners have the idea of the activity and which learners do not; i.e., make an informal diagnosis.)

1. "Who is in front of everyone?"

L: "Mary!"

I: "Right!"

BLUE   YELLOW   RED

FIGURE 3.1   Manipulative materials for exercises in relational terms

2. "Joe is in front of _____ ."

(Ask the learners to complete the statement to make it true.)

L: "Beth!"

I: "Good!"

Ask the three people to face toward the other side of the room. (Now Mary is last in line, Joe is next, and Beth is first.)

Give the same question and incomplete statement as you did when they were facing in the opposite direction.

If the learners understand, discuss who is *behind* the line, in a similar manner. If the learners do not understand, continue with other examples and demonstrate.

After understanding has been demonstrated by the learners, four people can be asked to get in a line. (Suppose Bob becomes number four.) Ask such questions as, "Who is in front of Bob but behind Joe?"

One step removed from this activity is the use of sets of similar objects that have identifiable fronts and backs, such as toy cars of different types or colors or animal figures of different kinds. Using such objects gives the instructor much greater flexibility. He or she can arrange and rearrange the objects in an order, discuss this order, and ask the learners for a variety of behaviors, which can be used to determine whether the learners do or do not understand the concepts being presented.

SUGGESTED ACTIVITIES

ACTIVITY

Alternative One. Arrange three toy cars of different colors in a row, as in Figure 3.1.

1. Ask the learner to put his or her blue, red, and yellow cars in the same order as you have them, with the red car in front of the other two.
2. Ask the learner to point to the car at the front of the line, behind the red car, or at the back of the line. The learner identifies the relative position of the cars.
3. Ask the learner to tell you which car is at the front of the line, behind the yellow car, or in the middle of the line.

The learner responds verbally to the questions. The development can be continued using pictures.

Alternative Two. Present pictures of objects arranged in an order, as in Figure 3.2. Ask the learner to copy the arrangement with animal figures. Make a statement or draw pictures in response to the relative position of an object. The instructor can expand this development by drawing pictures of objects in a certain order and asking the learners to respond as they did when dealing with the objects. Another alternative is to state where specific objects are located and ask the learner to place the objects in relative position to each other.

FIGURE 3.2   Display appropriate for exercises in relational terms

These activities are clearly designed for young learners. The substitution of different materials results in an activity that probably takes place daily in an automobile repair shop; for example, the procedure used in changing an oil filter involves the relative position of the filter, the cover, and the fasteners.

Through informal diagnostic techniques—by asking questions and observing results—the instructor should be able to determine the appropriate alternative to use. The teacher should present the vocabulary involving relative position when it is discovered that the learners possess these concepts but not the vocabulary with which to express them.

The ability of the learners to know and use the words and phrases in Table 3.1 should be diagnosed and taught, if necessary, prior to other activities in geometry.

Descriptive words.   Words and phrases involving relative size are descriptive words of geometry, and they usually come in opposite pairs: Large–small; big–little; long–short; tall–short; fat–thin; wide–narrow. Such descriptors are relative, since a softball is big when compared to a golf ball but small when compared to a basketball. Relative size seems to be one of the easier concepts of geometry, but it should be handled in as much detail as relative position if necessary. Relative size foreshadows the concept of similarity, which appears later in the content of geometry.

Words that serve to describe conditions of real objects represent the part of geometry that is the most flexible, the part that has the least restrictions imposed upon it. That geometry is called *topology*. (This type of geometry is named here only to provide a reference point for the reader; the term need not be used with learners.) This type of geometry has been called preschooler's geometry; each of the diagrams in Figure 3.3 represents the same idea to a young child, and in terms of topology they are the same.

One of the concepts of topology that is very important with respect to the content of geometry has already been outlined in the previous sections: order, or order constancy (with the people, cars, and animal figures). Flexibility is the general rule, but the order is maintained, for example, if three sections of ribbon, each section a different color, are placed in order and sewed together

TABLE 3.1   Language of geometry: Samples of "relative position"

Word or phrase	Example
in front of	in front of the TV
at the front of	at the front of the gas line
in back of	in back of the train
behind	behind the building
beside	beside the flagpole
to the right of	to the right of the chalkboard
to the left of	to the left of my home
between*	between the bookends

* *Between* is a very important relative-position word and should be developed in as much detail as possible to produce understanding on the part of the learners. Activities similar to those for *in front of* should be used.

Space, Relations, and Figures

FIGURE 3.3   Shapes that are topologically equivalent

as red, blue, yellow, as in Figure 3.4. It doesn't matter how we twist or turn the ribbon; the order, reading from the red end, will be red-blue-yellow. Related to this is another concept of topology, *betweenness*. The blue piece of ribbon remains between the red and yellow pieces.

Another important concept in this area is *open-closed*. Most preschool children have had some experience with these terms, so the tasks of the instructor are to extend and to refine. Again, demonstration becomes the teaching procedure. Open the door. Perform the activity and state what you are doing. Close the door. Perform the activity and state what you are doing. Ask a learner to open the door. Perform the activity and state what you are doing. Ask the learner to open the door. Ask a learner to open a window. Demonstrate, if necessary. Ask a learner to close the door. Ask a learner to close the window. Demonstrate, if necessary.

To continue the development, three or four boxes, canisters, jars, and other containers with lids should be obtained. All lids should be removed from the containers, and the learners should be told that all of the containers are now open. Objects will fall out of the containers if they are turned over. All of the lids should be placed on the containers, and the learners should be told that the containers are now closed. As long as the lids remain on the containers, objects will not fall from them.

A C T I V I T Y

SAMPLE INSTRUCTIONAL ACTIVITY

Select two containers of each kind; open some of the containers and close the others. Distribute containers to the learners. Take one of the open containers and close it by putting the lid in place. Ask the learners to close their containers.

Take one of the closed containers and open it by removing the lid. Ask the learners to open their containers.

Ask the learners to identify another closed container in your group of containers, and another open container. Assess whether or not the concept of *open-closed* is understood by the learners.

The development can be continued using the following input behavior by the instructor and output behaviors of the learners.

Instructor displays pictures of open and closed containers.	Learner opens or closes his or her container.
	Learner identifies which containers are open and which are closed.

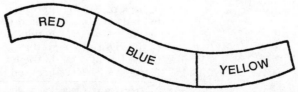

FIGURE 3.4   Starting point for manipulative exercises in order constancy

The words *on* and *off* should be emphasized as the lids are put in place or removed. Since this concept of *open-closed* is prerequisite for descriptive words of relative position, it is necessary that the learners develop a good understanding of it.

Descriptive words of relative position.   To foreshadow such terms as *side, face, area,* or *volume,* it becomes necessary to describe relative position using such terms as *inside, outside,* and *on.* Before this topic is introduced the learners should have some idea of relative position; they should be able to demonstrate or tell where objects are in relation to other objects. In addition, the learners should have an idea of the *open-closed* concept. The ideas of *inside, outside,* and *on* can be introduced early in a mathematics program, but they are often overlooked. Include them when needed—early elementary, middle, or even junior high school.

The concepts of *inside, outside,* and *on* extend from the idea of a closed container. In order for the learner to *see* models of these conditions, transparent containers, such as jars, should be used.

A C T I V I T Y

SAMPLE INSTRUCTIONAL ACTIVITY

Present three models for the learners.

1. A jar with a block (or other object) inside, with the lid in place. (See Figure 3.5.)
   Ask the learners to take a container and place an object inside of it. Make sure a lid is in place. Note: An object could be placed in a shoebox without a lid, but to be accurate, a lid should be used. Perhaps a classroom pet could be used to show that it could get out of the box if the lid is not in place. To teach topology, a lid must be included.
2. A block outside a jar, with the lid in place.
   Ask the learners to take a container and place an object outside of it.
3. A block on a jar (actually on the lid) with the lid in place.
   Ask the learners to take a container and place an object on it.

From this activity the learners can identify the condition, describe the condition, or draw the condition.

Different containers or pictures of containers may be used to produce the desired behaviors on the part of the learners.

At this point in the geometry content development, the terminology has been given geometric meaning. Where specific objects are located in relation

FIGURE 3.5   Examples of inside, outside, and on

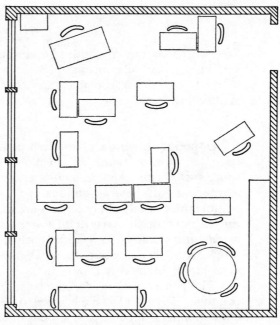

FIGURE 3.6   Diagram of a classroom

to other objects has been defined. *Relative position* in three-space has been given some meaning; a natural extension of the learner's world has occurred. Movement within three-space is also of interest in the nature of topology. We are only restricted by physical barriers and our ability. Flexibility is again the rule; we should not always move in straight paths, because learners do not move in this manner.

Changing relative position and paths.   Activities of a child involve such things as walking, running, hopping, skipping, and riding, all of which take place along some *path:* sidewalk, road, street, highway. These activities should be used to develop more content of geometry. The idea of going from one place to another is a basic concept of geometry because of the path that is followed, and the idea relates directly to what learners do naturally. The concepts of *open* and *closed* are revisited here and modified to include kinds of paths.

As we conclude this initial development of the preschooler's geometry, it should be noted that the youngster's knowledge and understanding of geometry may be greater than his or her understanding of number concepts. With this in mind, the teacher should consider elements of geometry as the *first kind of content presented to early elementary school children.*

ACTIVITY

An activity that can be used to introduce the concept of *path* is the game of Follow the Leader. (This is an appropriate activity for early elementary school children.) A leader should be elected. Give specific directions that everyone is to follow the same path that the leader takes. A diagram of the classroom should be drawn on the board, complete with desks, as in Figure 3.6.

ACTIVITY

SAMPLE INSTRUCTIONAL ACTIVITY

One of the first tasks is for the learner to determine *kinds of paths.* Ask a learner (say, Joe) to start at his or her desk and walk to the door. Draw this path on the chalkboard.

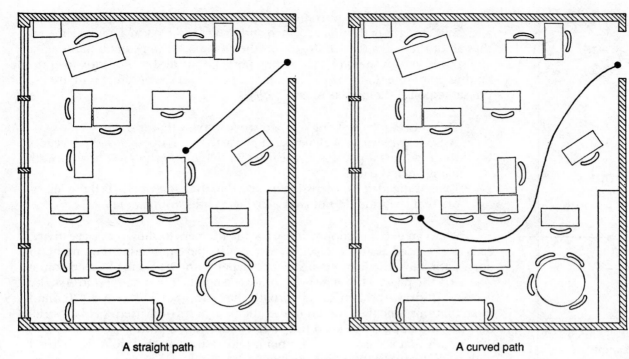

A straight path   A curved path

FIGURE 3.7   Examples of paths

Mention that Joe's desk was a starting point, the door was a stopping point, and Joe's path was from his desk to the door. Ask another class participant to walk along a different path from Joe's desk to the door. Draw this path on the chalkboard.

Give everyone a copy of a diagram of the classroom. Help each learner to identify his or her desk and the door on the diagram. Ask each learner to draw a path from his or her desk to the door. Check to determine whether learners can do this successfully. Assist, if necessary. Have the learners draw three different paths from their desks to the door.

Specific words of geometry can be introduced here by asking such questions as the following and drawing the given answers on the chalkboard. Emphasize starting point and ending point.

1. Who could walk in a straight path from his or her desk to the door? Select some likely candidate, use a straightedge (ruler, yard-, or meter-stick) and draw a correct response on the chalkboard. State, "This is a straight path."
2. Who could walk in a curved path from his or her desk to the door? Select some individual with a correct diagram and draw his or her response on the chalkboard, as in Figure 3.7. (Use different-colored chalk if possible.)

A C T I V I T Y **SUGGESTED ACTIVITIES**

If all of the learners understand the ideas of straight and curved, the activity can be extended to include wavy paths or broken line paths.

Activities on the geoboard could be used to reinforce these concepts. Routes on maps could be used to show paths from one location to another.

Several geometric concepts are being introduced with these activities—points, lines, line segments, straight, curved, movement of an object from one position to another (a transformation).

After the learners have an understanding of straight and curved, the idea

**FIGURE 3.8
Example of
an open path**

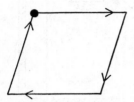

**FIGURE 3.9
Example of
closed path**

of open path should be introduced. The instructor can demonstrate, show, and draw straight paths that are open and describe the path. Tell the learners that all of the paths that you drew on the chalkboard were open paths since you started at a given point, the starting point (at the desk), and stopped at the ending point (the door). Since the starting point and the ending point are not the same point, the path is open.

- Demonstrate by walking along a straight path that is open.
- Ask a learner to walk along a straight path that is open. Ask the learners to draw such a path. Ask a learner to tell another person how to walk along such a path.
- Extend the idea by discussing a curved path that is open. Tell the learners that they should not *cross over* their paths at this time.

Here is an excellent opportunity for the learners to show some creativity. This can be used at any level with learners who have an understanding of an open path. Give the learners a sheet of paper with two points located somewhere on the page. They may even locate the points if they suggest it. Ask the learners to draw the fanciest open path they can. (See Figure 3.8.) Remind them that this time they should not let the path cross over itself. (This restriction is being imposed to keep the path a simple open path.)

A path that starts at a certain point and returns to that point is a closed path. (See Figure 3.9.) This idea should be developed in as much detail as open; perhaps more, since it is the basis of polygons.

Such ideas as round trips should be discussed; for example, from home to school to home. As much effort as possible should be made to relate these ideas of open and closed paths to what the learners actually do, since these ideas are really the first formal, specific concepts of geometry.

Conclusion of initial development. The concepts of this section of specific words of geometry—relative position, descriptive words, relative size, open and closed paths—begin to serve the two goals stated at the beginning. With an understanding of these concepts, the learners can approach the specific elements of geometry with a better chance of success and with a more positive attitude.

Since the preceding material can be considered prerequisite for other content in geometry, learners at any level of development should possess these concepts before going to a more formal development. If the development of content in geometry has been followed since the time the youngster entered school, open and closed paths should be familiar to early elementary school learners (K–3). However, this development may have been delayed because of poor sequence in geometric content and activities, some disability on the part of the learner, or the absence of such material in the program. Whatever the reason, an instructor should determine where the learners are in the development and continue from that point. Some tenth-grade learners may not have the concepts of *open* and *closed* and thus need an introduction to these concepts.

We can now discuss a sequence for geometry. The geometry of the K–6 grades should begin in three-dimensional space with the study of solid shapes, since, from the earliest age, the child's experience is with solids. Characteristics and parts of three-dimensional objects can be used to reinforce such concepts as relative position. Two-dimensional objects and their characteristics and parts should be emphasized in grades K–2. The families of polygons, circles, point-line relationships, and angles should represent the geometric content for grades 3–6. Properties of two-dimensional figures, transformations, and simple proofs should outline geometry for a junior high school. Axiomatic, deductive geometry follows at the senior high school level.

Geometry Outline

To preview the remaining parts of this chapter, a developmental outline is presented for grades K–8.

Three-dimensional objects
    Characteristics of three-dimensional objects
        Round (ball-sphere)
        Flat (box-cube; rectangular solid)
        Round *and* flat (can-cylinder; cone)
    Parts of three-dimensional objects
        Tops
        Bottoms (bases)
        Sides
        Faces
        Edges
        Corners (vertices)
        Tracing faces (simple closed paths)
Two-dimensional objects
    Square (from box)
        Line segments (straight portions of lines)
        Points (intersections)
        Inside, outside, on concepts
        Square shape and square region
    Rectangle (from shoe or cracker box)
        Line segments
        Points
        Inside, outside, on concepts
        Rectangular shape and rectangular region
    Circle (from can)
        Closed curve
        Inside, outside, on concepts
        Circle shape and circular region
    Circle (from cone)
    Triangle (from prism)
        Line segments
        Points
        Inside, outside, on concepts
        Triangle shape and triangular region
Point-line relationships
    Collinear and noncollinear points
    Intersecting lines

Parallel lines and line segments
    Square and rectangle
    Parallelogram
    Rhombus
    Trapezoid
Perpendicular lines and line segments
    Square and rectangle
Diagonals
Symmetry
Other parts of lines and their relationships
    Rays
    Angles
        Right angles
        Acute angles
        Obtuse angles
    Angles in polygons
Relationships in geometry
    Congruence
    Similarity
Motions in the plane
    Translations (slides)
    Rotations (turns)
    Reflections (flips)

DESCRIBING GEOMETRIC FIGURES

Introduction

The outline of content represents a more complete approach to geometry than appears in most textbooks and contains concepts appropriate for learners with special needs. For example, the important relative-position concept of *betweenness* is not listed in the scope and sequence chart of some programs. The same is true for paths. In addition, many series have been written under the assumption that the learners have certain preschool experiences. For example, experience with three-dimensional figures is assumed, and such ideas as similarity and congruence are introduced together on just two pages. Does this provide a good background for the learner? We could find many such examples of this kind of development, which creates mathematical gaps for all but the "above average" learner. To avoid this dilemma the suggested activities must be modified to include more options for both the learner and the instructor. They should be limited to only one or two concepts. Often such modifications are suggested in the teacher's guide, but they must be built into the program to make them functional.

One example will serve to illustrate how this modification procedure can be placed in the Interactive Unit format. One workbook page (Figure 3.10) contains pictures of a triangular solid and a cube at the top of the page, with a triangle and square traced on the respective figures.

The remaining portion of the page contains figures of triangles, squares, rectangles, and general quadrilaterals, with the directions to color the inside of each triangle green and the inside of each square yellow. Within this worksheet are the implied relationships between three-dimensional and two-di-

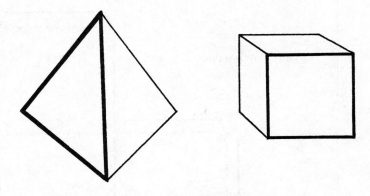

Color the inside of each triangle green.
Color the inside of each rectangle yellow.

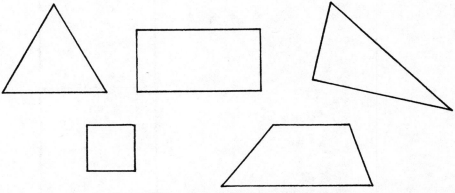

FIGURE 3.10 Sample workbook page

mensional figures, the *inside* concept, and the sorting of polygons. Unless understanding of each of these concepts has been achieved, how will the learner be able to complete this worksheet with understanding? If a learner already has a handicap in this area, you can imagine the possible results—confusion and development of a poor attitude toward mathematics. In addition, the input modalities of most programs are say and write, and the output modalities are write or say (sometimes). What if these modalities and interactions do not match the learning styles of the learner?

To modify this work sheet requires the following type of procedures:

1. Develop the concepts of *inside, outside,* and *on* in relation to each other, by using three-dimensional figures.
2. Relate three-dimensional figures to two-dimensional figures by having the youngsters draw polygon patterns.
3. Develop the *inside, outside,* and *on* concepts of two-dimensional figures.

As can be seen, multiple-concept activities must have as a prerequisite the complete development of each independent concept. Some learners with handicaps to learning and achievement may not be able to handle the multiple concepts but will do very well with each separate idea.

Descriptive Geometric Characteristics

Given the idea that the study of mathematics should begin with spatial characteristics, a view of objects in three-space seems appropriate. Again, a

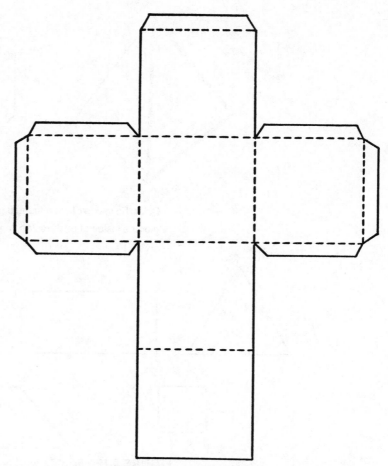

FIGURE 3.11 Pattern for a cube

look at complete (or all of the) objects will be considered. Toys can play a major role here. A ball (sphere) can be seen and touched. It will roll. A ball is round. Using familiar objects, the geometric vocabulary of learners can be extended, and intuitive ideas can be developed. For example, a ball is free to roll in many directions, but a can (cylinder) is restricted to two directions unless it is turned, and a box (cube) is difficult to roll in any direction.

With the material in the section on initial development as prerequisite material, the descriptive word list can be extended. Learners can be asked to roll a ball along a *straight line path,* and the ball will roll because it is round. A basketball can be used to continue the development of *inside, outside,* and *on* by deflating the ball and then inflating it again. *Inside* becomes an intuitive concept, since the learners cannot see the inside of the ball. A football can be used to demonstrate rolling along a curved path.

A box (perhaps a box that is used to contain a corsage, since it approximates a cube) can be used to continue the *inside-outside-on* idea, but now attention should be called to the sides of the box and the top and bottom of the box. Examples of words and phrases that should be used are "The box rests on its base and bottom," and "The lid goes on top of the box." Learners should be asked to trace the shape of the base of the box by placing the box on a piece of paper and drawing around it. Perhaps this should be the first school experience with a square. If each side and the top are traced, labeled, and placed together, a pattern for a cube is formed, like the one represented in Figure 3.11.

Activities involving these solids are suitable for early elementary school and should be developed as group activities. The strategy to be used with the three-dimensional figures involves the following steps: (1) introduce the complete figure, (2) describe the general characteristics of the figure, such as flat or round, and (3) intuitively introduce the parts of the figures—the sides, top, base, and edges. The input modalities (behaviors) used by the teacher are display and state. The output modalities (learner behaviors) are identify and say. The teacher displays the figures, and learners identify the figures and the parts. The teacher describes the characteristics of the figures, and learners describe and identify the figures.

Activities such as this can be used to check degree of motor coordination and perception. We should not assume that this is such an easy activity that everyone can do it.

The edges and corners (vertices) may be identified if the learners understand the prerequisite concepts. The names cube and rectangular solid need not be identified formally until square and rectangle have been named. Note that the figures produced by tracing the bases of the figures are the square and the rectangle. The bases themselves are square regions and rectangular regions.

Here we are introducing parts and part-whole relationships. Individuals who function in a left-brain manner only or who are slower in development may experience difficulty here. Instructors must be precise in verbalizing the characteristics and in showing what parts are involved. If difficulties persist, lessons in this content area may have to be delayed. However, they should not be forgotten, or a significant mathematical gap will be created. Once such a gap exists it may be "covered up" as the learner tries to compensate by avoiding the issue or giving alternative answers. For example, to continue to call a vertex a *corner* will result in difficulties with angles and their measure.

Different strategies for presenting descriptive geometric characteristics are represented by the following sample instructional activity.

ACTIVITY

SAMPLE INSTRUCTIONAL ACTIVITY

I: Present models of rectangular prisms—boxes, wooden blocks, or cubes—to the learner.
   Ask the learner to point to several examples of classroom objects like the ones displayed.
L: Learner identifies the objects.
I: Ask the learner to describe each figure in terms of the shapes of its sides.
L: Learner states whether sides are square or rectangles.
I: Provide the learner with diagrams (patterns) of a cube and a rectangular prism. (See Figures 3.11 and 3.12.)
   Ask the learner to fold the patterns to make them look like the models presented.
L: Learner constructs cube and rectangular prism (assist learners if necessary).

Attention should now be called to the number of sides or *faces* (a new term) of each figure. Pyramids, cones, and spheres should be introduced only as three-dimensional figures of the real world at the early elementary school level. They can be named and described at the junior high school level, and properties can be defined in the senior high school courses. We foreshadow volume by introducing the ideas early.

To move from three-dimensional figures to two-dimensional figures, do activities such as the following.

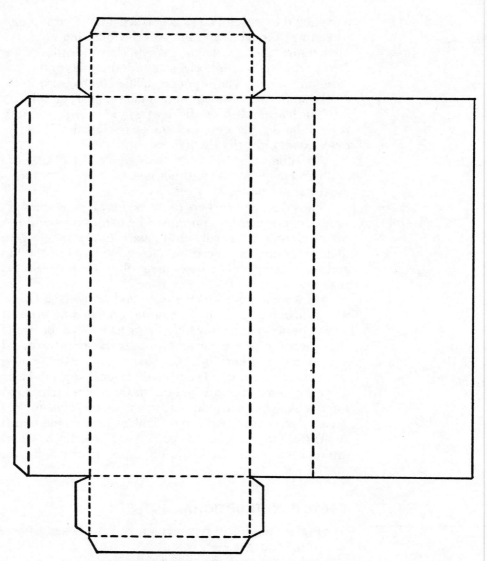

FIGURE 3.12 Pattern for a rectangular prism

SAMPLE INSTRUCTIONAL ACTIVITY

I: Present models of cubes and rectangular solids to the learners. Ask the learners to use the models and trace the bases onto a piece of paper, producing the square and the rectangle.

L: Learner constructs the square and the rectangle.

As the learners trace the bases, a relationship can be implied that the figure formed is a closed path: you start at a point and return to the same point without crossing the path.

To develop the idea of a circle, a can (cylinder) should be used. The side of the can is round and the top and bottom (base) are flat. The idea of a surface being flat is a significant idea in the development of geometry in a place (or flat surface). The sides of a box, as well as the top and bottom, are all flat surfaces.

This concept is more difficult than it appears to be, and it is essential if the concepts of Euclidean geometry are to be developed. The learner is being asked to "see" elements and ideas on a flat surface, which may be horizontal,

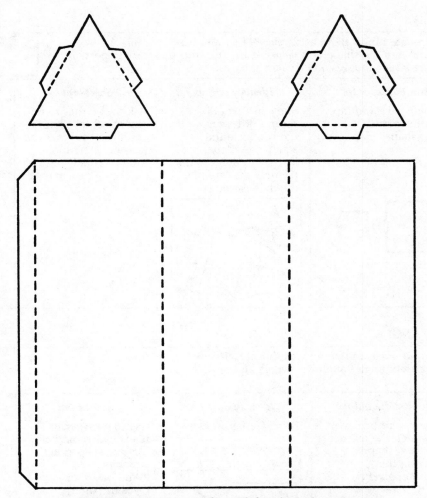

FIGURE 3.13   Pattern for a right triangular prism

vertical, or slanting. Placing each object on each of its own flat surfaces may serve to reinforce the idea. Several examples should be given, such as walls, floors, ceilings, desktops, the surface of the water in a glass, and a sheet of paper.

<table>
<tr><td>A C T I V I T Y</td><td></td></tr>
</table>

SAMPLE INSTRUCTIONAL ACTIVITY

I: Display examples of cylinders (tin cans of different sizes or film canisters) for the learners.

　　Ask the learner to find classroom objects that are like the ones displayed.

L: Learner identifies objects.

I: Ask the learner to describe the objects shown. Encourage the learner to note the circular bases.

L: Learner describes cylinder.

To produce the triangle, a right triangular prism should be used. "Slicing" a right triangular prism so that the slice is parallel to the base will result in a triangle. An actual glass prism from a science laboratory would be helpful here. Again, a pattern to fold together to make the prism would also be helpful. Such a diagram is outlined in Figure 3.13.

Space, Relations, and Figures

SQUARE

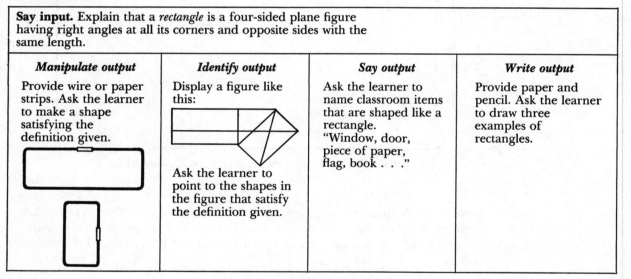

Manipulate input. Bend a 12″ piece of copper wire or similar material into the shape of a *square*. Tape the ends together. Repeat, making larger squares and smaller squares.

Manipulate output	*Identify output*	*Say output*	*Write output*
Provide wire and tape. Ask the learner to make shapes like yours.	Show pictures of several different figures, including squares. Ask the learner to point to the pictures that are like the models shown.	Ask the learner to tell you about the shapes you made. Prompt the learner to mention sides of equal length and square corners.	Provide paper and pencil. Ask the learner to draw pictures of the shapes you made.

RECTANGLE

Say input. Explain that a *rectangle* is a four-sided plane figure having right angles at all its corners and opposite sides with the same length.

Manipulate output	*Identify output*	*Say output*	*Write output*
Provide wire or paper strips. Ask the learner to make a shape satisfying the definition given.	Display a figure like this: Ask the learner to point to the shapes in the figure that satisfy the definition given.	Ask the learner to name classroom items that are shaped like a rectangle. "Window, door, piece of paper, flag, book . . ."	Provide paper and pencil. Ask the learner to draw three examples of rectangles.

Figure 3.14 continues

Since these ideas appear early in the school program, it would *not* be desirable to ask the learners to construct such patterns. Such activity can become very meaningful later, at the junior high school level.

For learners with special needs, geometry at this level can be used as a change of pace, a base for creativity, and, most important of all, an opportunity to experience success. They can relate the three-dimensional figures to real-world objects, and models can be built to see and touch. Exploration and discovery are functions that can be served by geometry. These learners, perhaps for the first time, have some flexibility in a mathematics program.

Shape

Attention to specific characteristics of three-dimensional figures leads to the study of shape, the third element in the study of (relative) position, (rela-

CIRCLE

Display input. Show several *hoops* of different sizes: wire hoops, plastic hoops, circular bracelets, etc.

Manipulate output	Identify output	Say output	Write output
Provide some modeling clay or wire. Ask the learner to make shapes like those shown.	Provide an activity sheet showing several different geometric shapes. Ask the learner to point out those shapes that are like the ones shown.	Ask the learner to describe the shapes shown.	Ask the learner to draw a picture of the shapes shown.

TRIANGLE

Write input. Write on the chalkboard:

A triangle is a closed plane figure having only three sides.

Manipulate output	Identify output	Say output	Write output
Provide wire and tape. Ask the learner to make a representation of the shape described on the board.	Show several models of polyhedra. Ask the learner to point to those shapes that satisfy what is written on the board.	Ask the learner to state the name of the shape that satisfies the definition of the board.	Ask the learner to draw a picture of the shape that satisfies the properties listed on the board.

FIGURE 3.14   Activities calling for the names of shapes

tive) size, and shape. The preceding information should be prerequisite material for any detailed study of shape. The learners should be asked to identify objects whose parts have the shape of a square, a triangle, a rectangle, or a circle. Sample instructional activities to show the possible interactions and developmental activities for the square, rectangle, circle, and triangle are described in Figure 3.14.

It should be noted that each of these activities has been selected to demonstrate different kinds of teacher input and learner output; they are not complete in that they do not contain all possible input and output behaviors for each figure.

As we leave this section, we change from a descriptive approach to an analytical approach, giving more attention to detail. A certain level of maturity is necessary here, because the models from the real world, which represent two-dimensional figures, require more insight to see the figures. For example, to see the bottom of a cracker box as outlining a square requires some thought once the box has been removed.

Two-Dimensional Figures

By tracing a figure to produce a square, a rectangle, a triangle, or a circle, the learner is drawing a simple closed figure. These closed paths become two-dimensional figures, and the learners should begin to describe these figures in regard to how many sides and corners (vertices) each figure has. The concept of measurement should be foreshadowed by noting how many sides of each figure have the same size. Such activities as developing a family of four-sided figures should be completed. A mobile can be built for the classroom to demonstrate the family. Figure 3.15 is an example of such a mobile.

A teacher could complete this construction and add the names of the various shapes as they are introduced in class. Later, the learners will discover the relationship among the members of the family. Pictures of each shape

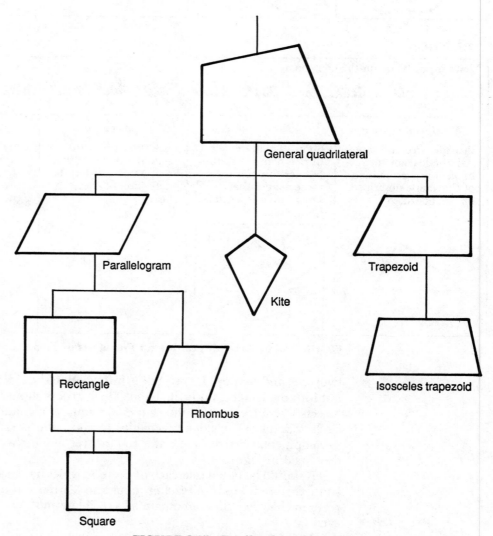

FIGURE 3.15   Family of quadrilaterals

should be identified in real objects. For example, ask the learners to find magazine pictures that have squares as part of the picture. The instructor should make an effort to find squares of many different sizes, to introduce the idea of similarity, and squares in various positions, to help the learner in divergent thinking. We can extend this to produce a problem-solving activity by asking and helping the learners to build as many different size squares on a geoboard as possible. Figure 3.16 contains some examples. From activities on the geoboard the learner can develop the ideas of points and line segments. The inside parts of the figures should be called *regions* to introduce the background material for area. All simple, closed figures in a plane have an inside and an outside.

A rectangle should be introduced by putting two squares together to show the learners that two sides of a rectangle are longer than the other two sides. We want the learners to distinguish between rectangles and squares. To accomplish this, we can have learners sort models, use templates to draw squares and rectangles, or label pictures as either squares or rectangles.

A family of triangles contains scalene triangles (no sides of the same length), isosceles triangles (at least two sides of the same length), and equilateral triangles (all sides of the same length). We should *not* use these names during the learner's initial development of the idea of the triangle, but the learners should see models and pictures of each type. They will learn in junior high school that triangles can be named in other ways, such as by angle size (right, acute, obtuse). We need to convince the learners that there are many different kinds and sizes of triangles. A good problem-solving activity is to establish the conditions for a triangle to exist and, at the same time, show the learners that not all situations in geometry have solutions.

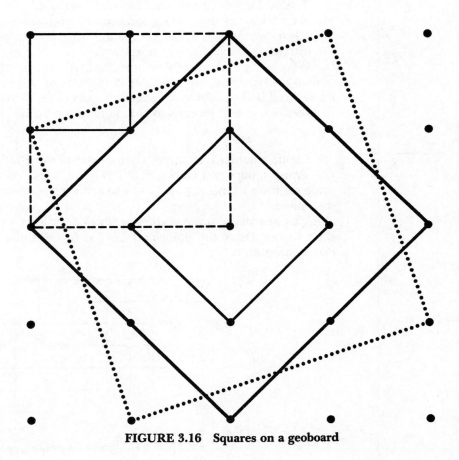

FIGURE 3.16 Squares on a geoboard

SAMPLE INSTRUCTIONAL ACTIVITY

Provide the learners with different sets of wood pieces.

- Set with lengths of 3, 4, and 5 units (inches)
- Set with lengths of 10, 8, and 6 units
- Set with lengths of 10, 10, and 5 units
- Set with lengths of 8, 8, and 8 units
- Set with lengths of 5, 5, and 10 units
- Set with lengths of 3, 4, and 9 units

FIGURE 3.17
Sample triangle

(It is important that we use material that will not bend.) Show the learner how to lay a set of three pieces together to form a triangle, as in Figure 3.17.

Ask the learners to make triangles by having the ends of each piece touch another end point. With the instructor's help, they should discover that the last two sets cannot be used to form a triangle because the two shorter pieces together are not as long as the longer piece.

To continue the activity, place all the pieces together and ask the learners to select sets of three pieces that will make a triangle.

Points, Lines, and Planes

We have discussed the ideas of relative position and implied that relative position can change by moving the object or figure along a path. This provides us with excellent opportunities to introduce other ideas of geometry.

To develop geometry of a plane and to provide a structure to the development, each learner should be given centimeter graph paper on which to draw. This development should be completed with the teacher demonstrating and the learner copying the demonstration.

What we are about to demonstrate involves several geometric concepts, and an instructor should realize that only one or two new concepts should be introduced and reinforced per mathematics class. This development ordinarily takes weeks and months to complete in the actual classroom.

Locate a point in the upper right corner of the graph paper on a corner of one graph square and label it *A*, as in Figure 3.18. Ask the learners to do the same on their graph paper. Check to see if all learners have completed this step successfully. The point has relative position. We want this point to move along a particular path. Ask the learner to draw a path as the point moves left four spaces. Draw the point and label it *B*. Continue by drawing the path down three units.

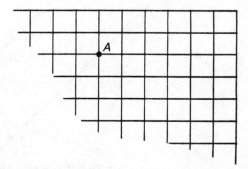

FIGURE 3.18 Point on graph paper

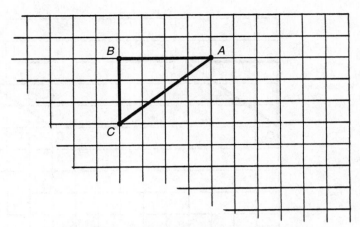

FIGURE 3.19 Triangle *ABC* on graph paper

Draw the point and label it *C*. Now, draw the path from *C* to *A*. We have a simple closed path made of three locations of the points and three line segments: a triangle. (See Figure 3.19.) Complete the other figures (polygons) in a similar manner.

Now we want to move the triangle (or other figures). Ask the learners to trace their triangles and cut a pattern for what they have traced. The traced, cut-out figure and the drawing are the same size and shape; therefore, we have introduced, informally, the concept of congruence. Explain to the learners that we want to move the triangle so that side *AB* is still from right to left (in a horizontal position) and side *BC* is still in the down-up (vertical) position. Let's slide the cut-out figure so that point *A* moves 10 spaces left. We discover that points *B* and *C* also move left 10 spaces. Trace the figure on the graph paper and label its points A_1, B_1, and C_1 respectively, as in Figure 3.20. Now let's slide the cut-out figure down from its new position (A_1, B_1, C_1) twelve spaces (Figure 3.21).

We discover that our triangle is still the same size and that all points of the triangle moved twelve spaces. Every point of the triangle followed paths that remained the same distance apart. Point *B* remained four units from point *A*. These paths are like railroad tracks in that they do not meet each other even if we continue to move along them. Such paths are *parallel*. If we change from one path to another, the two paths must meet or *intersect*. We can see our

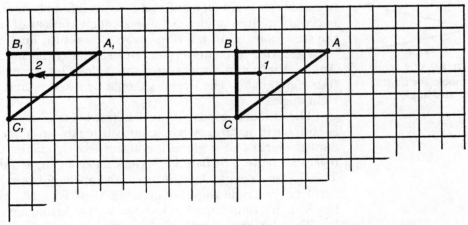

FIGURE 3.20 Moving triangle *ABC* from position 1 to position 2

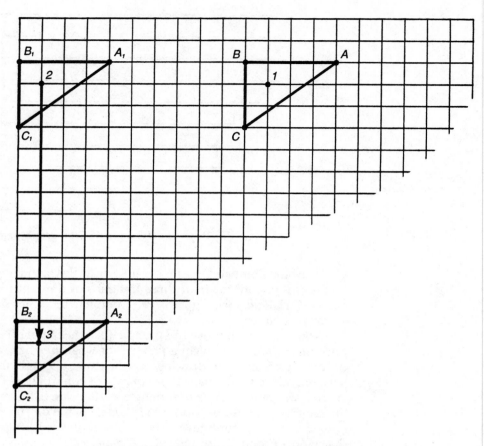

FIGURE 3.21 Moving triangle *ABC* from position 2 to position 3

graph paper as parallel paths going up and down or across. When the up-and-down paths cross the paths that are across the page, they intersect. Some streets cross each other or intersect.

We should note that sides of our closed figures intersect to help us produce the figure. We should also note that figures like the square and rectangle have opposite sides parallel.

The lines of our graph paper could be extended by adding other pieces of graph paper; we say that a straight line can be extended in two directions forever. We want to deal with parts of lines. One part of a line is a ray. A ray begins at one point on a line and extends infinitely in one direction. Rays are necessary to draw angles, because an angle consists of two rays coming from the same point. Figure 3.22 contains symbols for lines, rays, and angles.

Pieces of a carpenter's ruler can be used to make models of various angles. Angles that fit into corners are right angles. Our squares and rectangles have four right angles. The learners should be asked to find pictures that contain right angles.

The idea of angle should be illustrated in as many ways as possible, since it is abstract in nature. A bouncing ball makes an angle as it hits the floor. Sunlight makes an angle as it bounces from a mirror. The roof of a house makes an angle. The hands of a clock make an angle if they are not together. The size of an angle depends upon the opening between the rays. An angle smaller than a right angle but more than zero is called an *acute* angle. An angle larger than a right angle but less than a straight angle is called an *obtuse* angle.

With the development of the angle, we can introduce another motion in the plane, a rotation or turn through a given angle.

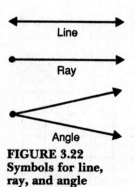

FIGURE 3.22 Symbols for line, ray, and angle

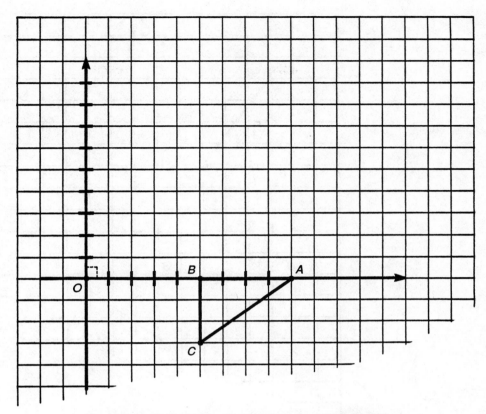

FIGURE 3.23 Triangle *ABC* on a coordinate graph

A C T I V I T Y
Start with the triangle that we produced in Figure 3.19. Copy this figure in the middle of a sheet of graph paper. We want to move the triangle from position 1 through a right angle to another position. The right angle for moving triangle *ABC* is already drawn in Figure 3.23. Point *B* is 5 units to the right of point *O*. Locate point B_1 5 units above point *O*. Point *A* is 9 units to the right of point *O*. Locate point *A* 9 units above *O*. Point *C* is 5 units to the right of point *O* and 3 units below line segment *OA*. Locate point *C* 5 units above point *O* and 3 units to the right of line segment *OA*. Figure 3.24 contains the answer.

Other figures should be turned through a right angle. If understanding is achieved, then angles of other sizes may be used.

We should now define congruent figures as figures in different positions that retain their exact shape and size. The tracing pattern for one will fit the other if they are congruent.

The learner should be asked to sort figures that are congruent from those that are not congruent, to identify congruent figures, and to draw congruent figures.

A C T I V I T Y
A reflection (or flip) may be used to introduce other geometric ideas, such as symmetry. Let's use our triangle (Figure 3.19) one more time. Through side *BC* draw a line, as in Figure 3.25.

Since points *B* and *C* are on the reflecting line, *f*, we leave them there. Point *A* is 4 units to the right of line *f*; therefore, we will locate A_2 4 units to the left of line *f*. Drawing the line segments A_2C and A_2B gives the triangle AA_2C in Figure 3.26. *A* has been reflected across line *f*. Line *f* becomes a line of

Space, Relations, and Figures

61

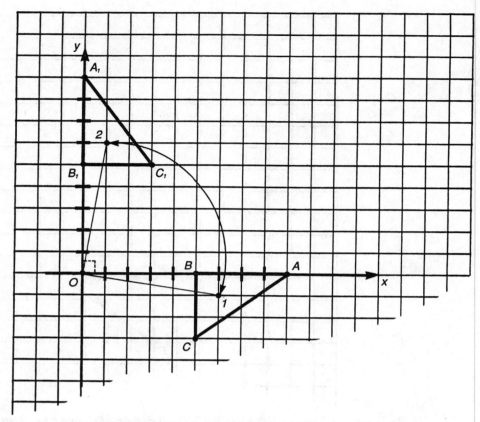

FIGURE 3.24   Moving triangle *ABC* from *x*-axis to *y*-axis

symmetry for triangle $A\ A_2C$, because the right side (triangle *ABC*) matches the left side (triangle A_2BC).

A line of symmetry is defined as a line in a figure that cuts a region into two congruent parts, which will then fold to fit the other exactly. Other figures may be checked by folding a paper model of them to determine how many lines of symmetry they have. For example, a square region has four different lines of symmetry, and a rectangular region has two different lines of symmetry.

SUMMARY AND CONCLUSIONS

In this development, from relative position to symmetry, we have attempted to outline appropriate geometric activities for the mildly handicapped learner in mathematics. If this set of material constitutes the total set of geometric concepts taught to and learned by the learners, an appropriate background in geometry has been established. Our desire is to have the reader expand upon our suggestions, enrich each topic, and modify what is usually presented in textbooks. To demonstrate what is possible, we conclude this chapter with a reference to microcomputers.

A Role for a Microcomputer

At this stage of the geometric development it would be appropriate and desirable to introduce the LOGO programming language. LOGO is an educa-

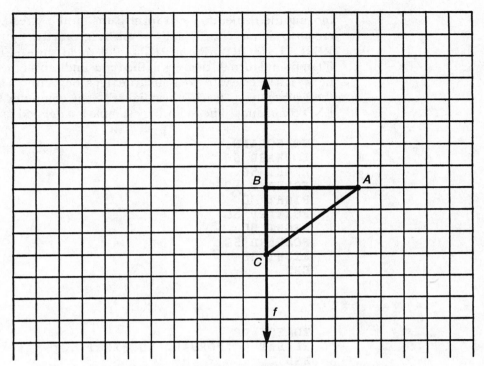

FIGURE 3.25 Triangle *ABC* with a line of reflection

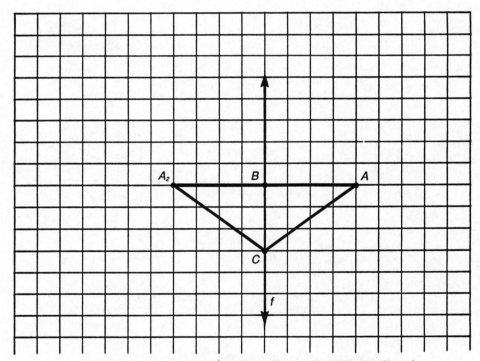

FIGURE 3.26 A reflection (flip) of triangle *ABC* over line *f*

tion-oriented language that allows the student to control the microcomputer instead of the microcomputer controlling everything the learner does. A "screen turtle" is controlled by the learner; he or she causes the turtle to move and turn by using simple words like FORWARD, BACK, RIGHT, and LEFT. A number is placed after the word to tell the turtle how many "turtle steps" to take or how many degrees to turn. After the learners have experimented by moving the turtle, they can form paths on the screen, even to the point of creating simple closed paths or plane geometric figures. The instructions REPEAT 4 [FORWARD 50 RIGHT 90] would cause the turtle to move forward 50 steps and turn 90 degrees to the right, and repeat this four times. The final figure is a square with sides 50 turtle steps long. After a figure has been tested, the learners can "teach" their figure to the computer by using the TO and END commands. The following example is a defined procedure for a square:

```
TO SQUARE
FORWARD 50
RIGHT 90
FORWARD 50
RIGHT 90
FORWARD 50
RIGHT 90
FORWARD 50
RIGHT 90
END
```

or

```
TO SQUARE
REPEAT 4 [FORWARD 50 RIGHT 90]
END
```

PRENUMBER CONTENT

INTRODUCTION

In addition to the research accomplished in the area of mathematics education, the increased number of young children experiencing preschool education of one form or another has led to greater awareness of the abilities of young children in the area of mathematics. Parents investing in the future of their offspring, while ensuring their welfare during the workday, are concerned that their children will gain the necessary steps on the path to success in school, an achievement that requires a large component of mathematics learning.

Parents of mildly handicapped children have no less interest in the overall achievement of their youngsters, and mildly handicapped children offer yet another reason for increased attention to the mathematics content proper to programs for young children. These youngsters spend a longer time experiencing this content. Hence prenumber activities deserve special attention from teachers of mildly handicapped children.

As is the case with many readiness activities, some prenumber experiences seem to have little connection to the arithmetic with which we are all so familiar. Nonetheless, research indicates that these activities indeed have some impact on the ease with which students later master number concepts. These topics include classification, patterning, and seriation.

CLASSIFICATION

Apart from its contribution to understanding numbers, the ability to classify enables us to deal with the wealth of stimuli bombarding the senses every minute of each and every day. To classify means to formulate a rule according to which one can sort the members of a group. For true classification to take place, the decision to use the rule or scheme for sorting the group of stimuli must come from within the person doing the sorting.

Sorting by Perceptual Attributes

Even very young children have a classification scheme working for them. The categories *strange* and *familiar* are adopted by children during infancy.

Presented with a set of small items, for example, raisins or candy, the two-year-old is likely to pick up a fistful or two. By heaping, the child has sorted from the rest of the pile what can be held in the hands.

At a later point in their development, youngsters seem to be able to impose a more rigid system of classification on a group of things. Presented with a set of shapes differing in color, size, and form, the three-year-old child may start placing the shapes in a row. For the first few items, a certain color may be the criterion for selection. The youngster might put a red square piece first, then a red triangular piece, then a red rectangular piece, and so on for the first five or six items in the row. After the sixth item is placed, the child might switch to shape as the criterion for selection. For example, if the sixth shape were a circular region, the next few shapes might also be circular regions, without any attention paid to color.

Accompanying this behavior is a tendency to sort objects in a pile by arranging them in a particular fashion. A three-year-old might take a deck of cards and arrange the cards in a row extending halfway across a room. Given a set of small plastic cars, that same child may line up the vehicles along the edge of a table. These and similar behaviors are important, for they represent the child's attempt to impose order on a number of things.

A later stage in this development more closely resembles what we think of as a sorting task. At this stage, given a set of shapes of different colors—red and yellow, for example—and asked to make two piles or groups, the four- or five-year-old will put some of the yellow pieces in one pile and some of the red pieces in another pile. Some red pieces and some yellow pieces, however, will probably still remain in the original pile. That is, the sort was accurate but not exhaustive. Only at a later point in the development is the sort both accurate and exhaustive, at which time the child has attained the skill sometimes called a *true sort*.

Sorting by Function

Until now, the examples given have involved perceptual attributes. Such attributes are clearly not the only ones available as criteria for sorting. Examine the following list of words: apple, bread, jelly, pear, cake, jam, plum, pie, peanut butter, grape, cookies, butter, pumpkin, noodle, mayonnaise, carrot, pudding, mustard, squash, jello, relish, potato, chair, tree, root, beet, table, sandwich, onion, desk, peach, bed, bureau. Take a paper and pencil and sort the words into groups so that the terms in each group have something in common. When you have completed the task, go back and try to sort the terms again, using a different rule.

Many different classifications for the terms in the list can be used.

1. Some people would separate the terms into two sets: food and non-food. Such a sort would be based on two large (in fact, all-embracing) categories.
2. Another sort might separate the terms into these groups: fruit, vegetables, things made with flour, spreads, things usually made with wood, desserts eaten with a spoon. Such a sort would be based on smaller and, in some cases, somewhat arbitrary classes or categories.
3. Still another sort might be based on the first letter of each term.
4. Another sort might base itself on the final letter of each term.
5. Still another sort might even be based on the number of syllables in each word.
6. Finally, the second sort described above might further subdivide the classes fruit and vegetables into those growing in trees and those growing on vines or as roots.

Sorting by Ideas

Notice that the efforts at classification described in the previous section have departed from the use of perceptual characteristics. Classification by function or role or previously formed concepts becomes the dominant rule for sorting. Thus, we have seen two types of sorting, the first by perceptual attributes, the second by concept or function. There is yet a third sort to be considered—sorting by ideas. Those readers who choose to outline the material in this book are performing this third type of sorting. They are grouping ideas under different headings.

The reader will observe that classification is a cognitive skill embracing more than efforts to deal with quantity. Language development itself is closely related to the ability to classify. Time spent helping mildly handicapped children who experience difficulty in learning to classify is an investment that will pay large dividends by way of greater success in attempting competence in other academic areas.

Sorting a pool of items according to a direction coming from another person is not a true classification task. To elicit classification, the teacher should simply ask the child to put items in groups so that all the items in each group have something in common. Directing a child to put all the red blocks in one container and all the green blocks in another might initiate a valuable learning experience for the child, but that learning is not necessarily classification. Such an activity might provide practice in recognizing red blocks and green blocks, or in putting things into containers, or in following directions. Those are valuable behaviors, but they do not represent the cognitive skill of classification.

In the sorting task assigned to the reader a few paragraphs back, after one sort was completed the reader was requested to go back and sort the terms again in a different fashion. In a limited way we were requiring some divergent thinking on the part of the reader. Mildly handicapped children need many opportunities to do things in different ways. Such experiences result not only in greater ease in divergent thinking, but also in the awareness that in many cases, more than one answer or response are correct.

A C T I V I T Y

SORTING ACTIVITIES

Some examples of sorting tasks are presented below. The reader is encouraged to do the tasks and then to develop one or two other classification tasks independently. In each activity the interactive combination from the Interactive Unit is noted.

1. Encourage the children to collect labels from canned goods, frozen food packages, and boxed foods. When a large enough assortment has been gathered, help the students to cut out and perhaps paste on cardboard the pictures of different food to be found on the labels. Shuffle the pictures and distribute them among the youngsters. Ask each child to sort his or her pictures so that all the pictures in a group have something in common. (Present/manipulate.) After each child has completed the sort, encourage him or her to explain the groups to the other students. (Display/state.)

2. Using the collection of pictures described in the previous activity, make several groups in which the members have something in common. Show your groups to the children. Ask them to describe other ways in which you could have sorted the pictures. (Display/state.)

3. This activity is described in terms of a set of cardboard or plastic pieces consisting of four different shapes of three different colors of two different sizes (attribute pieces). Such a set can be purchased from many educational supply companies. Without too much difficulty you can also make such a set,

with the children's help. Obtain six large sheets of sturdy cardboard. Using poster paint, paint two sheets yellow, two sheets red, and two sheets blue. From a yellow sheet cut a large circular region, a large nonsquare rectangular region, a large triangular region, and a large square region. From the second yellow sheet, cut smaller regions similar to the large regions. Repeat for the red sheets and the blue sheets.

Place the collection of attribute pieces in front of a group of students. Ask the children to sort all the pieces into *two* groups so that all the pieces in each group have something in common. (Display/manipulate.) When the sort has been completed, ask the students to name the attribute common to all the pieces in each group. Most likely, the youngsters will have sorted the pieces according to the attribute size. (Display/state.)

Repeat the activity, but this time ask the children to sort the set of attribute pieces into three groups. Once again, ask for the attribute on which the sort was made. Finally, ask the students to sort the pieces into four groups and to name the attribute used for the sort.

Discuss the influence of the size of each group on the number of differences in the attribute chosen. Return to the two-group sort. Ask the students to describe how each of the groups could be further subdivided. Encourage different responses. (State/state.)

4. Many household tasks require sorting. Provide the youngsters with a set of eating utensils of different kinds. Ask them to arrange the items so that desired pieces could be easily found. Other items for this type of activity might include laundry items, groceries, and dishes. (Display/manipulate.)

CLASSIFICATION AND SETS

Very often one can spot some confusion about the relationship between classification activities and working with sets in programs designed for young children or for mildly handicapped children. Certainly when a child sorts a collection of objects into smaller groups based on attributes perceived to be common to all the members of each group, he or she is building sets. The members of a set, however, need have no attributes in common other than that of belonging to the same set at the same time. Thus, any collection of objects is a set. Further, a set exists even where there are no members at all. We can talk about the set of female presidents of the United States, for example.

Because of the close relationship between sets, number meaning, and operations on numbers, students should have experience with sets. These activities can both precede and accompany instruction about numbers and the operations on numbers. Activities with sets should include experiences in comparing sets, joining sets, noting attributes common to the members of a set (when such attributes exist), and recognizing members common to more than one set at the same time. At no time is the use of the word *set* necessary. The skills are perhaps best described with examples. The activities that follow have been used successfully with mildly handicapped children.

COMPARING SETS

1. Before beginning this activity, put several pictures of sets in different places around the classroom. These sets should represent different numbers. Place enough pictures of sets so that there are two sets per child. Use any

pictures you have available, or help the students to make pictures of different sets prior to this activity.

Keep a picture showing a set of five objects or fewer for yourself. To begin the activity, divide the learners into two groups, the *greater than* group and the *fewer than* group. Show all children the picture of the set you have saved for yourself. Tell each member of the *greater than* group to find a picture of a set having more items than the set in your picture. Tell each member of the *fewer than* group to find a picture of a set having fewer items than the set in your picture. Each child is to bring his or her choice to you for approval. As the children make a correct choice, tell them to look for another picture satisfying the same requirement. Give the youngsters a time limit—for example, ten minutes. At the end of that time, announce that the search is over. Use the pictures the children have found to set up a *greater than, fewer than* bulletin board. (Display/identify.)

2. Clear a space in the classroom. Ask the children to give you their full attention and to listen closely. Clap your hands twice. Ask the students to clap their hands more times than you clapped yours. Repeat the task several times, varying the number of claps. You can change the activity by performing another action, such as jumping or hopping, and by requesting the children to make fewer movements than you do. You may also appoint a child to act as the leader of the activity. (Manipulate/manipulate.)

3. Arrange a group of several students in a row facing the rest of the class. Give each child a picture representing a set. The sets should be such that different numbers are represented. As the leader, you construct a set of three members. Ask a child who is not holding a picture of a set to identify, by walking up to and tapping on the shoulder, a child who is holding a picture that has fewer items than the set you made. A correct identification allows the selector to take the place of the child originally holding the picture. Vary the activity by asking the children to identify sets that show a greater number of items than the one you have made. You may also wish to appoint a child to act as leader. (Display/identify.)

JOINING SETS

1. Ask three boys having brown eyes to stand as a group. Ask four girls having brown eyes to stand as another group. Discuss the size of the the two groups and the fact that all the boys have brown eyes and all the girls have brown eyes. Then request the two groups to form a set of children having brown eyes. Discuss what the members of the new group have in common, and the size of the new group. Vary the activity by selecting other characteristics, or sets of different sizes. For each repetition, discuss the common attribute and the larger number found in the joining of the two smaller sets. (Display/state.)

2. Use three sets of blocks, or other small items, each of which represents a different number—for example, a set of three, a set of six, and a set of one. Tell the children to watch carefully as you join two of the sets. Join the two sets that represent the two largest numbers. Give the child three sets like yours. Ask the child to put those two sets together that will give the largest possible set. Vary the activity by requesting the smallest possible set. (State/manipulate.)

COMMON ATTRIBUTES

1. Material used for language development activities can be helpful in developing facility in recognizing attributes. Such materials would be sets of

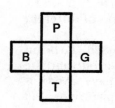

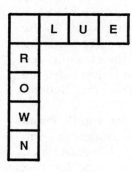

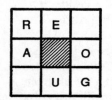

FIGURE 4.1
Word puzzles

cards on which are pictured animals of different kinds and different colors, flowers or other plants, household furniture, or other household items. Distribute the cards from such a set among the children. When you describe a group—all the members of which have the same one, two, or three attributes—the students holding the cards picturing the members of that group should stand. For example, if the cards distributed represented animals of different kinds, the group described might be the animals with short gray fur or the animals that are brown and fly. (State/identify.)

2. This same kind of activity can be extended to introductory work with numbers and letters. For example, the children might see a display like the following: A, 2, 7, D, M, W, H, G, O, 4, 9, 1. The youngsters would be asked to point to all the figures that have only straight lines. At a later time, the children would be asked to select the letters that have only curved lines in their configuration. (State/identify.)

3. Many software programs designed for young children provide experiences in recognizing one or two attributes. Their use can help to develop awareness of attributes.

4. Discuss with the children the different things that can be bought in a grocery store or in a supermarket. After several items have been named, tell the children that you are going to describe items by listing two groups to which that item could belong. They are to think of something that can be bought in a grocery store or a supermarket that would satisfy your description. For example, begin by saying, "I am thinking of something that is round and can be eaten. What am I thinking of?" Encourage the children to name several items satisfying your description. As the children become proficient in naming items, restrict the categories to which the items could belong. For example, "I am thinking of a round vegetable that is green." (State/state.)

The ability to recognize multiple attributes of something is closely related to the ability to recognize membership in more than one group at a time. This ability is important in language development as well as in mathematics.

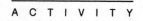

MEMBERSHIP IN MORE THAN ONE SET AT A TIME

1. On a sheet of paper, draw a set of red shapes and a set of circles. Ask the children to name or to draw something that could legitimately be found in either set. (Display/state, write.)

2. Show the children puzzles similar to those illustrated in Figure 4.1. Ask the children to use the choices given to complete the puzzles so that words are made in each direction. Vary the activity by presenting puzzles like those found in Figure 4.2. (Write/write.)

PATTERNING

Mathematics might be said to deal with the basic patterns that human beings find in (or impose upon) their environments. What, then, does "mathematics education" deal with? Presumably the way that human beings *think about, communicate about* or *learn to deal with* these basic patterns [Davis 1984, p. 2].

Many different patterns become familiar to children at a very young age. The movement as a parent rocks an infant back and forth in a rocking chair, the rhythm and the rhymes in nursery songs, the familiarity of stories heard over and over again, alliteration and internal rhyme in stories such as those of

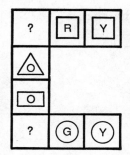

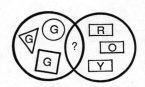

Red
Orange
Green
Yellow

**FIGURE 4.2
More puzzles**

Dr. Seuss—all of these have value largely because of repetition and the opportunities for rehearsal such repetition provides.

This same iteration acts as a basis of prediction, a skill that forms the foundation of many of our daily activities. Our interactions with other people are examples of patterns, and so are the cycle of the seasons, the designs evident in cross sections of certain fruits and vegetables, the distribution of genetic traits, and the life cycle itself. Without patterns, our existence would be chaotic. Without the expectancies we form on the basis of past experience, emotional stability would be impossible.

For mildly handicapped children, explicit instruction directed toward a search for patterns and the recognition of the patterns forming a part of their surroundings becomes an asset that more than justifies the formal inclusion of such content in curricula for these children. Because mathematics is the study of systematic patterns of relationships, the inclusion of formal instruction in patterns within the mathematics curriculum seems to be a most suitable strategy. An added bonus from this approach is the readiness for further work in quantitative processing that experience with patterns can provide.

The essence of a pattern is a repeating sequence of events. The events

Make each pattern longer by drawing pictures in the blank spaces.

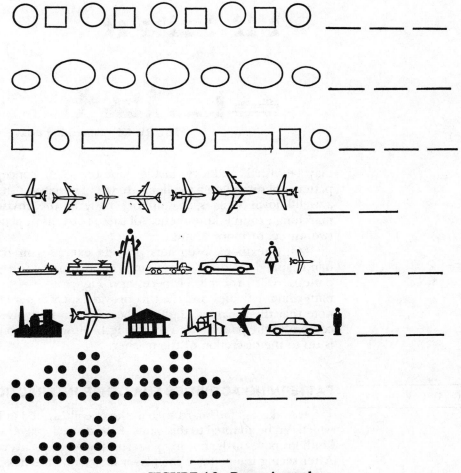

FIGURE 4.3   Patterning tasks

Prenumber Content

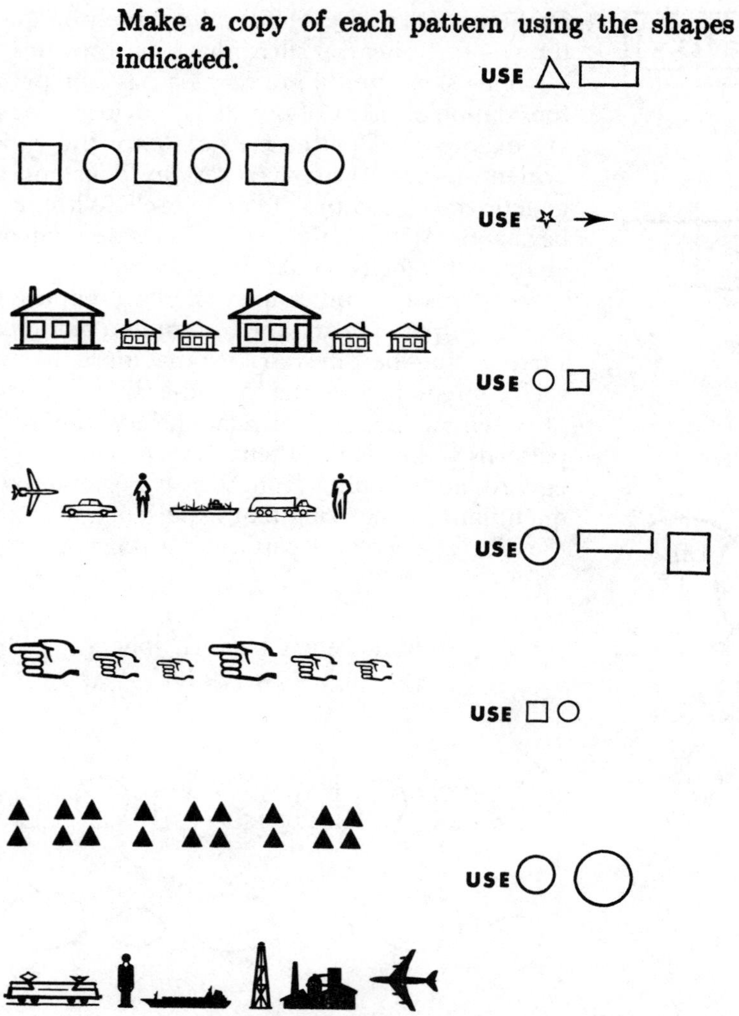

Make a copy of each pattern using the shapes indicated.

USE △ ▭

□ ○ □ ○ □ ○

USE ✧ ➤

USE ○ □

USE ○ ▭ □

USE □ ○

▲ ▲▲ ▲ ▲▲ ▲ ▲▲
▲ ▲▲ ▲ ▲▲ ▲ ▲▲

USE ○ ○

FIGURE 4.4 Patterning tasks

may be visual, auditory, tactile, kinesthetic, or conceptual in nature. The patterns observed on a tile floor are visual in nature; the recurring melody in a well-known song is an auditory pattern; the rhythm experienced by a marching group is kinesthetic. All these patterns depend on the repetition of two, three, or more stimuli.

Many textbook programs provide exercises in patterning. For mildly handicapped children, many more opportunities to recognize and produce patterns will probably be necessary. Figures 4.3–4.4 provide a sample of patterning activities designed to provide such experiences. The reader will note that the activities gradually become more involved with respect to the content of the patterns and their length. How far to go with a particular child is up to the discretion of the teacher.

A C T I V I T Y

PATTERNING ACTIVITIES FOR OTHER INTERACTION COMBINATIONS

1. Make a game board such as the one illustrated in Figure 4.5, or use one which can be adapted to this game. Provide each player with a marker of some kind. Prepare cards showing patterns. Show each player one pattern at a time. After seeing the pattern, the player is to name the next three elements. If he or she does so, that player moves a marker three spaces forward. Players who

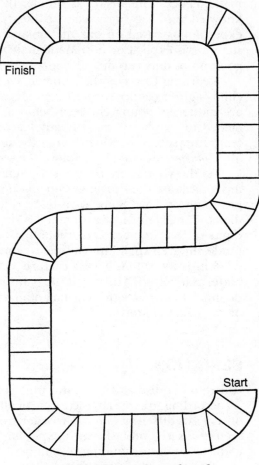

FIGURE 4.5   Game board

name only two successive items correctly move ahead two spaces. Players who name only one successive element move ahead only one space. The first player to reach the finish mark is the winner. (Display/state.)

2. If available, wallpaper samples offer good examples of visual patterns. Obtain several such samples, and discuss with the children the patterns that are used. Then assign each child the task of creating some wallpaper showing a definite pattern. (Display/write.)

3. Prepare a large display of a line or two of music with which the children are familiar. The song chosen should contain a recurring melody of fairly short duration. Point out the repeated sequences of notes. If possible, play the sequence for the children on a piano or on another instrument. Play a record of another song familiar to the children. Ask them to listen for repetition and to let you know when they hear it. (State/state, manipulate.)

4. Work with the children to create a gymnastics routine consisting of sequences of simple movements that are repeated. Encourage the children to create their own sequences after one routine has been learned. (Manipulate/manipulate.)

5. Read to the youngsters a poem or nursery rhyme in which rhythm plays an important role. After the first two verses, suggest that the students clap softly to highlight the rhythm. If the poem or rhyme is a short one, read it a second time while the children clap. (State/manipulate.)

Some readers may question the relationship of the above activities to mathematics. The answer is simple: the recognition of patterns demands a knowledge of the rule dictating the sequence that occurs. We build certain expectations based on our understanding of that rule, whether the understanding is explicit or not. Mathematics is replete with such rules, rules that govern not only certain behaviors but also the results of those behaviors.

Learning to recite the number names becomes an easier task when the child begins to notice recurring sounds. After the first thirteen number names are mastered, young children begin to notice the pattern of the "teen" attached to number names already learned. That they recognize and predict from the pattern is evident from the unsolicited "tenteen" that often comes after *nineteen* as children explore the sequence.

As the youngsters progress in their familiarity with numbers, other patterns facilitate the learning. Our place-value system is based on a pattern of grouping by tens. Some divisibility rules are based on patterns of sums of digits. For example, if a number is divisible by three, then the sum of the digits of that number is divisible by three. Kindergarten students and first-grade children spend much time writing the numbers from one to one hundred in rows of ten. That structure, correctly used, helps to emphasize the patterns involved. The use of the patterns inherent in our number system can do much to facilitate the learning of mathematics. That facilitation is the focus of the next chapter.

SERIATION

Allied to but distinct from patterning is the ability to order the members of a set according to some characteristic of all the members. The arranging of a set of sticks of different lengths in order from largest to smallest, or vice versa, is seen as a seriation task. Such experiences contribute greatly to the development of an understanding of number.

The process of ordering the members of a set according to a criterion requires several skills. Let us suppose for a moment that a child's task is to arrange three sticks of different lengths in order, from shortest to longest. The child must first identify the smallest piece by making all the comparisons possible. After that comes the identification of the shorter of the two remaining pieces and the placement of this piece next to the one already placed. Finally the youngster must place the remaining stick at the end of the sequence and then compare each stick with the one preceding and/or following it, to check that the sequence is correct.

Observe that even for this relatively small set of three objects, the child must use not only the prerequisite skill of comparing pairs of sticks, but also the new skill of comparing one stick with two other sticks, one on each side of it. Note also that with the addition of only one more stick to the set of three, the ordering task involves many more steps. Clearly, for many youngsters, seriation is not a skill that comes easily.

The connection between seriation and number concepts can perhaps be more completely illustrated by another example of a seriation task. A popular toy consists of a number of nesting cups, all having the same shape. The child delights in correctly placing the cups, one inside the other, until all have been properly fitted. It is also in this sense that seriation is related to number understanding. Three is not only greater in magnitude than two, which in turn is greater than one; in a real sense, *oneness* is contained in *twoness,* and *oneness* and *twoness* are contained in *threeness.*

A logical property characteristic of sets on which seriation can be imposed is transitivity. If we use the set of nesting cups mentioned previously as an example, we will note that if cup *b* fits inside cup *c,* and if cup *a* fits inside cup *b,* then of logical necessity cup *a* must fit inside cup *c.* Ideally, then, seriation activities help students not only to form better concepts of numbers and the relationships among them, but also to gain a better grasp of transitivity.

SERIATION TASKS

The content of seriation activities should consist of different dimensions along which quantity can be ordered. Thus, length, weight, size of area, number, pitch, and volume all provide the context for seriation tasks.

1. Provide the children with a set of bells, each of which sounds a different note. Ask them to order the bells according to note. (Say/manipulate.)

2. Show the students a random arrangement of sticks of different length. Ask them to work together to order the sticks from shortest to tallest. Observe carefully the strategies suggested and used. (Display/manipulate.)

3. Obtain a set of boxes of different sizes. In the boxes place items of different weights, taking care that there is no correspondence between the size of the box and the magnitude of the weight of the item placed therein. Ask the students to order the boxes according to size. Then ask them to order the boxes according to weight. (Display/manipulate.) Discuss with the students their experiences with the second task.

4. Show the children two sticks of different lengths. Give a child a third stick, the length of which is different from that of the first two sticks. Ask the child to place the third stick so that the three sticks are arranged in order according to length. Repeat the activity, each time varying the length of the third stick. (Display/manipulate.)

5. Display three sets of different cardinalities—for example, a set of three items, a set of seven items, and a set of ten items. Ask the youngsters to arrange the sets according to size. Observe the strategies adopted by the students. Do they attempt to order the sets without altering the arrangement of the members of the set? Do they line up the members of the sets to examine the magnitude of such an arrangement? Do they count the number of items in each set? (Display/manipulate.)

CONCLUSION

Although we have described the acts of understanding discussed in this chapter as prenumber concepts, the existence of these skills clearly does not cease with the attainment of number concepts. It seems likely that such understanding develops even further when children begin to deal with numbers on a formal basis. We will see their influence when we discuss the meaning of number in the next chapter. For mildly handicapped children, however, time spent nurturing the ability to sort, to recognize patterns, and to order members of a set is time very advantageously used.

WHOLE NUMBERS

INTRODUCTION

As a prelude to this chapter, we suggest that the reader look around and notice the many examples of numbers to be found in everyday surroundings. Our environment is replete with examples of numbers: license numbers, house numbers, the time of day, the date, one's weight, the price of beef, and so on. Your list is probably different, and that difference serves to emphasize the point being made: examples of numbers abound.

One other point deserves to be made. Read each of the following sentences:

No hay escritos en la pizarra.

Siki ya pili kufika makhadimu
Makatoka wedenda kukhudumika
kazi imezosalia.

Die sammklichen Vereinsgesetze der
deutschen Staaten enthatlen auf
Grund eines Beschlusses des seligen
Bondestages Folgende Bestimmung.

$$3 + 4 = 7$$

Relatively few people can read and understand all the sentences. On the other hand, the last sentence—the number sentence—probably has meaning for anyone who can read any one of the sentences above it. That relative universality of many, if not most, of the symbols in the language of mathematics forms the springboard for the following discussion.

Mathematics and children have long gone hand in hand. Young children take delight in repeating the number words over and over again. Children of most cultures learn to draw and eventually to write the symbols corresponding to those number words. In many countries, mathematics is one of the few subjects that follow students through a major portion of their schooling. Thus, the knowledge of at least some mathematics becomes part of the academic repertoire of most students, even those whose handicaps in learning pose special problems, for mathematics is the language we use to deal with quantity.

One can argue a case for the claim that the child's growing awareness of space, quantity, and number parallels that of the human race. Menninger (1977) suggested that our relationship to our environment necessitated our relationship with numbers. The evidence indicates that just as the child learns to count "one, two, three" long before he or she learns to write or to calculate, so did spoken number language precede the written. As the very young child counts, "one, two, many," so did human beings find it difficult at first to progress beyond the second number word.

The story of zero is another instance of the time needed by the human race to develop the number system as we know it today. Although the concept of *nothing* or *no more* becomes familiar to the child at an early age, as it did to people in the early stages of civilization, the knowledge of a symbol to denote this concept and the use of that same symbol to *hold a place* for a missing power of ten are signs of a high degree of sophistication relative to the understanding of number. The evidence seems to indicate that human beings arrived at such cleverness no sooner than the early middle ages (Menninger 1977). A longer time was required before mathematicians designed the concept of the *empty set* as being the set with numerousness zero. Yet there are those who expect the child to appreciate both of these interpretations of zero by age six or seven.

There are other instances of the need for cognitive maturity on the part of the race and on the part of the individual before certain concepts achieve a firm status. Examples are the development of an efficient and effective sequence of number words and symbols, the use of number to designate the size of a continuous entity (measurement), as well as to denote the number of items in a group (counting).

As the child's growth in the use of numbers reflects to some degree that of the race as a whole, so do some of the tactics a child adopts to help himself or herself along the way. Ginsburg (1986) pointed out the logic behind the number words used by the uninitiated, especially with reference to the words signifying decade changes: the use of "tenteen" for twenty and "tenty" for one hundred. Menninger (1977) cited the primitive use of bridging words by many cultures.

Many young children use their fingers to compute during the early stages of acquiring computational skills. Finger counting has had and still enjoys a prominent place in the heritage of many peoples. The "finger counting" that so many primary grade teachers work so hard to discourage seems to have its roots in a source much more powerful than the need to find an answer for an example on an arithmetic paper.

In light of the context briefly described above, it is clearly dangerous to cram too much formal arithmetic into the experience (academic or otherwise) of a young child. Humankind has trod a fascinating path on its journey to its present awareness of and skill with numbers. Young people should enjoy at least some of the same discoveries. Mildly handicapped children must not be hindered still further by programs that do not allow sufficient time for basic number concepts to develop. In this chapter we shall outline both content and methodology ranging from early number experiences through basic number concepts.

EARLY NUMBER EXPERIENCES

Determining just when young children become aware of numbers is difficult. Certainly most children can count to ten by the time they enter kindergarten.

Yet those who teach youngsters in kindergarten or first grade need little persuasion that such counting skills have little to do with understanding numbers.

Such comprehension seems to come at a later time, from experiences different from those needed to memorize the sequence of counting words. That is not to say, however, that the process of learning the sequence of number names is unimportant. We will examine the development of rote-counting skills, the learning of the meaning of number, and the behaviors needed for rational counting.

Rote Counting

The expression *rote learning* usually connotes memorization with little attention to meaning. Rote learning of the number words implies, then, the acquisition of the correct order of the number names one, two, three, and so on, without necessarily understanding the meaning of the arbitrary sequence or the individual words. Ginsburg (1986) pointed out, however, that only the first thirteen number names are totally arbitrary, at least in the English language. After those names the list becomes easier for children to learn, because they can formulate rules to help themselves in the counting endeavor. Thus, after completing the sequence *fourteen, fifteen, . . . nineteen* for the first time, many children will name the next number *tenteen.* After being introduced to *twenty* and *twenty-one,* most children can continue to *twenty-ten.* The learning of *thirty* marks attainment of the last nonconforming word; from that name on, the remaining words that name numbers less than one hundred follow rules already acquired by the learner.

Rote counting, whether aided by the formulation of rules or not, is a necessary skill for youngsters to possess. Without the correct sequence of the number names at their command, children will be unable to do any meaningful counting. Mildly handicapped learners will probably require a longer period of time to acquire the correct sequence of the arbitrary names. When rules dictating the sequence can be formulated, efforts to do so should be encouraged.

Ginsburg (1986) also cautioned that young children seem to limit the number of names they learn at any given time. Some children master one through five before moving on; others will attempt one through seven; others will accept a different challenge. The young mildly handicapped child may still be learning only a few new number words at a time during his or her primary grade experiences. Such a child will profit more from the experiences outlined in the previous chapters than from a lot of daily drill designed to teach rote counting.

Meaningful Counting

The behavior we call counting seems simple enough until we examine the different conditions under which it can take place. The things that we count are many and varied. Imagine each of the following groups, examining closely the tactics most adults would use to count the things under consideration:

- Pennies that have been saved in a jar.
- The cards in a deck.
- The change left in one's pocket at the end of the day.
- Beats while playing or listening to a piece of music.

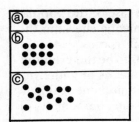

FIGURE 5.1
Counting tasks

- Someone's pulse.
- Jumps taken during a turn at jump rope.
- The different kinds of birds seen during a nature walk.
- The size of the population of the United States.
- The items in each frame in Figure 5.1.

Clearly, we use the simple behavior called counting in a variety of situations. Each group in the list requires some combination of behaviors different from the skills required by another group. The mature counter has mastered all the behaviors.

For example, in counting pennies from a jar, we can remove each penny from the pile as we count it. We can also pile the pennies in groups of ten to keep track of how many have already been counted. We usually count the cards in a deck by removing one card at a time from the deck and placing each card in another pile. The change left in one's pocket at the end of the day is usually such that counting by ones is unnecessary and undesirable. Further, the counting takes place by different increments: by 50, then 25, then 10, then 5, then 1, if coins of all values are included.

Counting beats, a pulse, or jumps requires a different approach. In the first instance, we tend to use part of the body such as the hand or the foot to note the beat, and count to the rhythm we feel. We seldom continue to count beyond a certain number—one, two, one, two; or one, two, three, one, two, three—but we repeat the sequence over and over again. Pulse counting demands counting a movement in another body while maintaining stillness in one's own. Further, unless great accuracy is demanded, we usually measure pulse for a fraction of a minute and multiply the result by the necessary number to obtain an estimate of the number of beats per minute. Counting jumps while skipping rope involves enumerating a motion initiated and experienced by oneself.

Counting the number of species of birds observed during a nature walk might be accomplished by holding in memory the last number named until a member of a different species is observed, repeating this procedure until the walk is over. A more sophisticated approach, however, uses tally marks or the name of each species observed, and a counting of the recordings after the walk is completed. Note the intermediary role of the record—one counts not the actual birds, but the tally marks or species names.

Counting the number of people who make up the population of a country represents yet another interpretation of counting. The number given as population size is not arrived at by enumerating all the people in the country. Nor need that number represent the size of the population at any given time. It is arrived at through a sampling procedure and a process of extrapolation. Nor is better accuracy necessary.

Finally, consider the different groups and arrangements illustrated in Figure 5.1. The items in the row in the first frame are easy to count. One moves one's eye or finger along the row, proceeding from one dot to the next. The members of the array in the second frame can also be counted by ones, but the mature counter would probably count the number in each row and multiply that number by the number of rows. The last group defies such an attempt. The lack of organization demands some kind of record-keeping strategy, perhaps that of crossing out each dot as it is counted.

The different kinds of groups not only require different counting techniques but also provide different challenges to checking the accuracy of the count. The count of the pennies, the change, the deck of cards, and the illustrations in the frames of Figure 5.1 can be verified by recounting. That is not possible for the jumps, or the pulse, or even the beats that have already passed in time. We can only count a second sample of each group or rely on a simultaneous count by someone or something else. In the case of the population tally or the bird-watching results, we can rely on someone checking the observations made and the procedures used for tallying purposes.

With such examples in mind, counting no longer seems to be the primitive task we sometimes consider it to be. It is a fundamental means of gathering data about the world of quantity that surrounds us. The attainment of skill in counting is a cognitive feat well worth attention. It is to the components of that skill that we now turn our attention.

One-to-One Correspondence, Equivalence, and Counting

Two sets, the members of which can be paired off so that no members remain in either set, are said to be in one-to-one correspondence. The ability to recognize this correspondence seems to be fundamental to the understanding of number. To be able to conserve number is to recognize one-to-one correspondence in spite of any dissimilarities in the arrangement of the members of each set. Many investigators have sought to determine precisely the relationship between the ability to conserve number and the understanding of number. The evidence seems to indicate that while conservation of number is an understanding mature counters have acquired, there exists no explicit hierarchy of the form: *conservation of number must precede understanding of numbers* (Resnick and Ford 1981; Lancy 1983). Hence tasks designed to facilitate the development of conservation of number can be included in the curriculum, hand in hand with counting activities.

Numerousness

leaf

FIGURE 5.2
One-to-one correspondence

By *numerousness* we mean the attribute that each of the following sets has in common with the others: the four fingers of a hand, the letters in the word *leaf*, the sides of a square, the tires on an automobile, the maximum number of dimes needed to make forty cents, and so on. If all those sets could be lined up, one under the other, an arrangement like that in Figure 5.2 would result. That understanding of one-to-one correspondence is the focus of this section.

The activities described below are designed to help mildly handicapped children understand numerousness and equivalence between sets. Gelman and Gallistel (1986) found that children can recognize the equivalence of two sets representing small numbers before they can recognize the equivalence of larger sets. For this reason we summoned many experiences with sets having numerousness between one and five before representations of larger numbers are attempted. For all sets and for all mildly handicapped children participating in these activities, we encourage you to require that the children demonstrate the equivalence or nonequivalence of two sets by pairing off the members of the groups. Each activity represents a sample of many tasks that can be developed along the same lines by changing only the materials used and/or the numbers represented.

A C T I V I T Y

UNDERSTANDING NUMEROUSNESS AND EQUIVALENCE BETWEEN SETS

1. Prepare a set of cards, perhaps sixteen in all, on which are represented sets: four cards showing two items on each card, four cards showing three items, four cards showing four items, and four cards showing five items. Distribute the cards in "hiding places" in the classroom. Display a set of four items. The children are to participate in a "treasure hunt" to find the cards that show sets that match the one on display. (Display/identify.)

2. Using raisins, carrots, or the like, and several saucers or small paper plates, prepare several small snacks consisting of two, three, four, or five raisins or carrots. Cover each dish with a paper cup. Show the children a snack consisting of three items. Ask them to guess which paper cup hides a snack having the same number of pieces. Allow the children to check their guesses until the correct plate is found. Repeat the activity until enough snacks have been found to enable everyone in the group to have a share. (Display/identify.)

3. This activity can take place over a period of several days. Use a Polaroid camera to take pictures of the children in groups of different sizes, so that the numbers from one to five are represented and each child is in at least two groups. Display the pictures so that each child can see all of them. After briefly discussing each photograph, tell the children that you are going to name a group of children. The children are to select the picture that shows *all the children* you name. Begin by naming groups of only one, two, or three children. When the children can handle that number, move up to four and five. When the children succeed in easily identifying the larger groups of children, move on to the second part of the activity.

For this task, tell the children that you will again list names, but the names will not be names of children. They may be names of flowers, animals, furniture, or the like. The children are to listen to your list and select a photograph that shows a group of children of the same size as the list you name. That is, if you list "cow, dog, cat, horse," the children are to select a picture showing four children. When beginning the activity, arrange the pictures so that the numbers are represented in order, one to five. When the children become skilled in identifying the correct photographs, rearrange the order of the photographs. (State/identify.)

4. Prepare a set of cards by writing one word on each card. Each word should have no more than five letters in it, and each number from one to five should be represented several times. Distribute the cards among the children, keeping a set of cards showing a word for each number from one to five. Show the students a card—for example, a card showing the word *bet*. Each child having a card on which there is a word having the same number of letters is to show that card. Repeat the activity for other words. (Write/write.)

5. When assigning partners for walks or other activities, distribute the cards made for the first activity in the section, making sure that each number is represented only twice. Each child is to find a partner by finding the child whose card shows a set equivalent to the one given to him or her. (Display/identify.)

6. In counting, one-to-one correspondence must be made between each item in the set to be counted and each number name recited. Children can be given such experiences in "tagging" (Gelman and Gallistel 1986) before they must use the number names. Show the child a picture of six items arranged in a row and move your finger across the row, stopping briefly at each item. Ask the child to do what you did. Observe carefully to ensure that the child stops at each item once and only once. Repeat the activity for sets of different sizes. (Manipulate/manipulate.)

Tell the children that you are going to recite a list of names. Each child is to make a mark on a piece of paper to represent each name you state. (State/write.)

Show the child a picture like that shown in the third frame of the illustration in Figure 5.1. Provide the child with a pencil. Ask him or her to mark each item in the picture once and only once. (Display/write.)

_____

Each of the foregoing tasks helps children to develop organization strategies. Such tactics play an important role not only in learning to count but also in most, if not all, mathematics content. Mildly handicapped children are often set back even further because they lack *modi operandi*—patterns of task performance that minimize the possibility of error due to confusion, distraction, or forgetfulness. Mildly handicapped students must be taught to develop and apply such strategies.

After learning to touch once and only once each item in a row of items, the child is ready to practice matching number names with touches. Before going directly to rows of objects, many children will profit from the experience of just counting taps. The teacher can provide some object, such as a book, to be tapped. The child is to touch the book and say "one" simultaneously. Each succeeding touch must be matched by the recitation of the next number name. After a number of taps matched by number names have occurred, the instructor can remove the book and ask the child to tell *how many taps in all* occurred before the book was removed. The exercise will provide practice in matching number names with touches, and in recalling the last number named to tell how many touches there were altogether. Once these two behaviors have been assimilated, practice in counting the objects in a row by touching each object and reciting the appropriate number can be undertaken. After each row is counted, the learner should be required to name the number of items in the row by repeating the last number name that was said. This drill can be repeated many times for groups of different sizes. Other groups can be introduced:

1. Items arranged in arrays.
2. Pictures of things arranged in a row or a column. Note that for these groups the child cannot physically move each item as he or she counts it. Other strategies have to be used. For example, a child could use a crayon to mark each item in some way as it is counted.
3. Pictures of arrays.
4. Groups of objects randomly arranged. Note that in this case, the counter is almost forced to move an item as it is counted, since no pattern exists in the arrangement.
5. Pictures of a group of objects randomly arranged.
6. Repeated actions, done by others or done by self. For example, as one child hops, another child counts the number of hops, or a child might count as he or she takes steps.
7. Repeated noises: the chimes of a church bell, fire whistles, and so on.

These activities are meant to be experienced over a long period of time. Little by little, children will acquire the basic elements of counting: the ability to recite the number names in the correct order, the ability to count each member of a set once and only once, and the recognition that the last number named in the counting process is the numerousness of the set. With these insights should come two other very important ones—the understanding that number is distinct from that which is being counted, and the realization that the order in which the members of a group are counted does not affect the outcome. We now turn our attention to these last concepts.

INTERPRETATIONS OF WHOLE NUMBERS

There are two major approaches to the development of number meaning. The first of these considers the cardinal property of a set as the vehicle of understanding and application. The other position argues for the place of measurement for representing numbers and the relationships between and among them. We shall consider each of these briefly, in turn.

Cardinal Property

We observed previously that children must be able to recognize the equivalence of two sets that are equivalent, no matter what the arrangement of the members of the sets is. This skill paves the way for an understanding of number as the cardinal property of a set. Number, then, becomes an attribute of a group of things. It is in this context that we can speak of twoness, fiveness, tenness, and so on. The number of items in a group is the last number named when counting the members of the set correctly.

An important feature of this approach to number is the following: a set corresponding to any given cardinal property contains within it subsets having cardinal properties corresponding to each of the numbers recited before the number of the set. That is, a set of four items has subsets containing zero, one, two, and three items. Hence the need for experiences with seriation before working with numbers as such.

Critics of the cardinal property interpretation of numbers cite the difficulty of representing large numbers as cardinal properties and the need for students to deal with negative numbers and imaginary numbers later on in their academic careers. For these critics, number is a theoretical construct that happens to find application in counting and naming the cardinal property of a set. For most mildly handicapped children, the concept of numbers as elements of a field that embraces many different forms of numbers may be inappropriate. For these children, activities like the following should serve as a powerful vehicle for introducing and developing the meaning of number.

A C T I V I T Y

UNDERSTANDING AND APPLYING NUMBER MEANINGS

1. Show the children two groups of items. The groups should differ in the nature of the members of the group, in the number of items in the group, or both. For example, in one set there might be three pieces of fruit, in the other three pencils. Discuss with the children the difference between the two groups (one contains fruit, the other pencils) and the similarity between the groups (both have three members). Repeat the activity for groups of other items and other numbers. (Display/state.)

2. Provide each child with pencil, crayons, and paper. Write a number on the chalkboard. Ask the children to draw a set having that many members. Check each child's work. Then ask the children to use the sets they have drawn to create another picture in which the original drawings are parts combined into a larger figure (Write/write.)

3. Provide each child with a set of small items, such as pennies. Arrange six pennies in a row. Ask the children to do what you did. (Manipulate/manipulate.) Count the pennies together, ending with the statement, "There are six pennies in a row." (Display/state.) Then rearrange the pennies in a 2 × 3 array, while the children watch. Ask them to do the same. Ask the children to predict the result of counting the pennies again. If some of the children seem unsure, count the pennies, again noting, "There are six pennies in all." Repeat the activity for different arrangements of the pennies, including random arrangements. During succeeding days, repeat the activity for other numbers. As a final evaluation, using only one group of small items, arrange ten items in a row and count them with the children. Then ask the children how many items there would be if you arranged the items differently.

4. Display several sets of items, some sets arranged in one row, others in arrays, others in different configurations. Count the members of a set arranged in a row with the children, going from left to right. Ask the children what would happen if you started counting from the right. If some students seem unsure, count the members again, this time from right to left. Note that

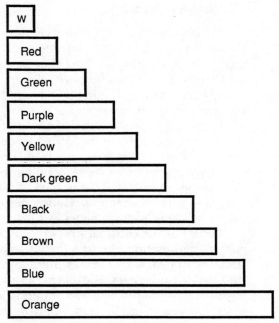

FIGURE 5.3 Cuisenaire rods

the number is the same. Repeat the activity for other sets, each time counting from different starting points. For an evaluation, show the youngsters a set of ten items in a row. Count the items with the children, going from right to left. Then point to the leftmost item and ask, "What will be the last number we say if we count from here?" The children should respond with *ten*. Then point to an item within the row and ask what would happen if you started from there and counted all the items in the row. Again the answer should come readily— *ten*.

Measurement

The measurement approach to number emphasizes the relationships between and among numbers. From this perspective, a number such as four is defined in terms of its role with respect to other numbers. Thus, four is twice two, four times one, one-half of eight, and so on.

The things used to represent number in this fashion usually depend on some physical attribute to represent the relationships. Cuisenaire rods, illustrated in Figure 5.3, represent certain relationships by way of length and color. Counting plays a relatively minor role in establishing these relationships, except insofar as that process is used to establish a standard to begin with, and then to determine certain relationships with respect to that standard. That is, three *two-rods* might be placed next to a *six-rod* to illustrate the relationship "six is three times two."

Proponents of this approach to number emphasize the manipulative nature of the materials used. Cuisenaire rods, for example, are used in developing concepts of different groups of numbers—for example, fractions and decimals, as well as whole numbers. Because relationships are the dominant focus of instruction, advocates maintain that the abstract nature of number becomes apparent relatively early in a child's experience with number, and that the students are not tied into a consistent representation of any given number.

In our work, we have used the concept of cardinal property to teach the meaning of number. Our decision to do so was based on our level of comfort with that approach and on our belief that the *manyness* of a set is a concept that is easier for young children to acquire. Many users of the measurement approach would argue strongly for the advantages such an approach offers. We suggest that in the long run, an instructor's expertise in using either approach will have the greatest effect on how successfully either method is used in any given situation. We do recommend, however, that some approach be used consistently, and that this approach provide for many manipulative experiences throughout the development of number concepts.

Ordinal Numbers

We use numbers not only to describe a given quantity, but also to designate position in a sequence. Numbers used for that purpose are called *ordinal* numbers. Children need to be as familiar with them as they are with the counting numbers.

The contexts within which ordinal numbers are used are many: position in a linear sequence (first in line, second in line, and so forth), the day of the month, relative standings in a given hierarchy (first vice-president, second vice-president, and the like), and others. These situations dictate which ordinal numbers need to be mastered. Aside from the days in a month, we most often use first, second, third, and last. That list, however, illustrates the need for formal instruction in the ordinal number names; the words *first, second,* and *third* bear little, if any, resemblance to their cardinal number counterparts. Aside from the spelling and the reading of these names, other concepts and skills require attention.

A major concept is the understanding of orientation or direction and its influence on ordinality. The person who is first in a line of individuals facing in one direction becomes last in the line if the people in line turn to face in the opposite direction. This concept can be built very simply, as the following activity illustrates.

A C T I V I T Y

INFLUENCE OF ORIENTATION OR DIRECTION

Using a bright lamp and large white sheets of paper, draw a silhouetted profile of each child. Ask each child to carefully cut out his or her profile. Arrange each profile with the chalkboard as a background so that all the profiles are facing in one direction. Ask the children to name the first person in the line of profiles, the second person, and so on to the last person. After discussing the lineup using ordinal number names, ask each child to turn his or her profile so that it is facing in the opposite direction. Once again, discuss each child's position. Note that when we count with cardinal numbers, position in the group does not dictate which number we attach to the item when we count. However, when we use ordinal numbers, position does matter.

Mildly handicapped children need to be able to place objects in an order representing ordinality; they also need to recognize objects in named ordinal positions. That is, when faced with a sequence of items, the child should be able to answer such questions as *Which item is first? Second? Last?* These skills can be developed daily through informal practice.

Writing the names of the ordinal positions is a skill that requires the learning of two separate systems of symbols. Ordinal number names are seen as words only in some contexts (for example, Our first speaker today is . . .), and as a combination of numerals and letters in other contexts (1st, 2nd, and so on). Using ordinal numbers successfully requires that the user be proficient in translating from ordinal numbers to cardinal numbers and vice versa. Thus, the mildly handicapped child must know that both *one* and *first* represent the starting point for a sequence of numbers, and that *one* is translated to *first* when order is important.

Place Value

Mathematics educators are in general agreement that understanding the base-ten nature of our decimal system and facility in using the written nota-

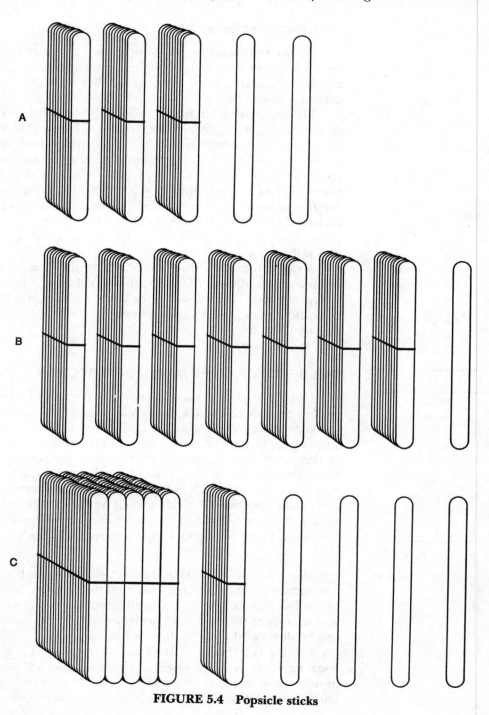

FIGURE 5.4   Popsicle sticks

tion representing this system are of maximum importance if real progress in the arithmetic of the rational number is to take place. Yet the best procedures for teaching place value and its relationship to the algorithms we use are far from determined. We can be quite certain, however, that mildly handicapped children will benefit from programs designed to build a solid understanding of our number system and the symbols we use for representing its organization.

Many different kinds of materials have been designed to provide manipulative experiences with the concept of ten as a grouping principle. We will describe some of these materials, attempting to highlight their principal advantages and, in some cases, to note cautions that should be observed in the use of the materials. The set of materials used in any given situation depends on the curriculum adopted by the school and/or the level of comfort of the teacher when using the materials.

A simple and inexpensive mode of representation makes use of popsicle sticks, tongue depressors, or coffee stirrers. Each stick represents one unit. When counting the sticks, the child is encouraged to group them into sets of ten, usually securing the groups with a rubber band. When ten groups of ten have been counted, these groups can be put together to make one group of one hundred. The use of the sets to represent numbers is illustrated in Figure 5.4.

The advantages of such a set of materials lie in their availability and relatively small cost. The disadvantages can be found in the cumbersomeness of their use for representing large quantities. For some children, however, that very cumbersomeness may provide an asset—that of realizing the magnitude of large numbers.

Popsicle sticks are also used in a system of representation that includes dried beans that are somewhat flat—for example, kidney beans. Each bean represents one unit; ten beans glued to a popsicle stick represent a group of ten; ten sticks joined together represent one hundred. The sticks are joined by gluing them to a sheet of paper or piece of cardboard. The use of the materials to represent numbers is illustrated in Figure 5.5.

The size of the unit (a kidney bean) results in a less cumbersome set of materials. The layout of the representations for ten and one hundred can be more easily tied to the powers of ten and their role in each new grouping. A unit is 10^0, and the bean represents a point having theoretically *zero* dimensions; a ten is 10^1, and the stick represents a line having *one* dimension, length;

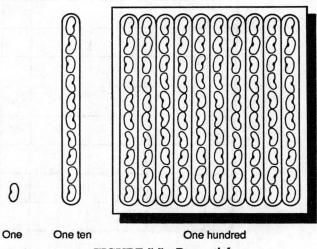

One   One ten   One hundred

FIGURE 5.5 Bean sticks

Whole Numbers

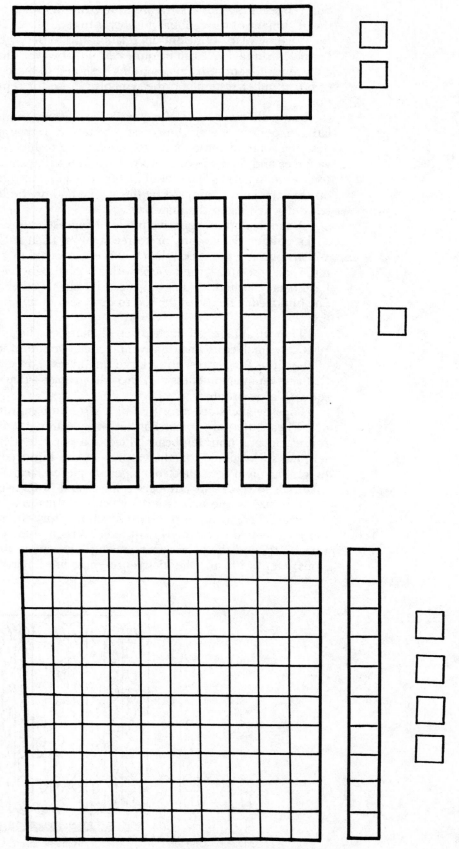

FIGURE 5.6   Base-ten blocks

one hundred is 10^2, and the block of sticks represents a square having *two* dimensions.

Unfortunately, this mode of representation does not easily lend itself to the illustration of one thousand, or 10^3. An added disadvantage is the frequency with which the beans can fall off the sticks, so that care must be maintained to ensure that each stick does in fact represent ten.

Materials that avoid either of the disadvantages mentioned in the preceding paragraph are Dienes' blocks and base-ten blocks. Dienes' blocks can be used to represent number systems with bases other than ten as well as ten; base-ten blocks assume a base-ten number system. In base-ten blocks, a centimeter cube represents one unit; a stick that is one centimeter by one centimeter by ten centimeters represents ten; a block that is ten centimeters by ten centimeters by one centimeter represents one hundred. Illustrated in Figure 5.6 (in reduced form), the base-ten blocks can also be used to represent one thousand through the use of the cube ten centimeters by ten centimeters by ten centimeters. Units are marked in the ten-stick, the hundred-flat, and the thousand-cube.

Another advantage of the base-ten blocks is the ease with which number representation can be transferred to graph paper made with squares of one centimeter on a side. Although place value is principally a symbolic concept, interactions of different types should be used in the development of the understanding of the meaning of the symbols. Unfortunately, base-ten blocks are somewhat expensive to buy. Schools often have only three or four sets, shared among several teachers. If at all possible, mildly handicapped children, as well as other children, should have the materials available at all times when working with numbers.

Similar to the base-ten blocks in one respect are the Montessori materials (1917) for developing place-value concepts. Small beads (usually gold in color) represent units; ten beads strung on a piece of wire represent tens; ten wires connected to form a block represent hundreds. With these blocks the children use color-coded cards on which the numerals are written. The children arrange the cards and the beads to represent different numbers. Such a representation is illustrated in Figure 5.7.

The Montessori materials share both the advantages and disadvantages of the base-ten blocks. In the Montessori materials, the color coding of the numerals does serve to highlight the different values of each digit in the written form of a number. Eventually, of course, the coloring must be deemphasized and value attached not to a color but to a place in the right-to-left sequence.

Until now the materials described reflect to a great degree the relative value of the powers of ten represented. Other types of manipulatives approach a higher level of abstraction. Instead of representing multiples of ten visually, these materials make use of size, color, or position to denote the digits of numbers. Plastic chips of different sizes, for example, can be used to represent units, tens, hundreds, thousands, and so on. Also, chips of different colors, with each color representing a different power of ten, can represent

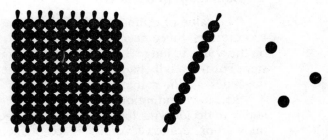

FIGURE 5.7   Montessori place-value materials

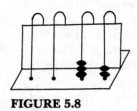

**FIGURE 5.8
An abacus**

numbers. In both these cases we are not restricted to representing only units, ten, hundreds, and thousands by the materials. Plastic chips are readily obtainable and relatively inexpensive. We must recognize, however, that the definitions we impose on the chips are fairly arbitrary. Such arbitrary definitions may not be as powerful in developing an understanding of a base-ten number system.

Related to the use of chips is the use of an abacus (Figure 5.8). Consisting of counters strung on wires, both ends of which are attached to a board, an abacus serves to represent numbers through counters brought forward on each wire.

Abacuses present several advantages. Since it is easily made, an abacus can be available to each child. It provides few difficulties as far as the management of the material is concerned. In addition to the cue provided by the right-to-left placement of the wires, color coding can be used to enhance the awareness of each place in a number. The size of the counters could also differ, providing yet another visual help. However, an abacus does require that the child learn the rules governing the use of this intermediary material.

The choice of which materials to use for place-value instruction is an important one. As we shall see in the next chapter, models for concepts should be such that children can grow with them. As is the case with toys, the best manipulatives are those that can be used over and over again, across many different levels of a topic and across many different topics. Mildly handicapped children should not have to switch from one set of materials to another each time a new concept is to be developed. Intimacy with a set of materials will enable a child to explore further applications of the materials and will enhance the discovery process. Teachers too, as they become more familiar with the potential of a set of materials, will find their level of enthusiasm and creativity expanding. At that point both teacher and learners become active participants in the instructional process.

We cannot leave the topic of place value without examining the question of scope. What degree of facility should mildly handicapped students attain in reading and writing large numbers? We recommend that much time and effort be devoted to the manipulative representation, reading, and writing of three-place numbers. Once these skills have been mastered, progression to numbers with four, five, or six digits requires the knowledge that the comma says *thousand,* but we read the groups before the comma just as we did before. That is, given the number 35,425, we read it as thirty-five *thousand,* four hundred twenty-five. Some children enjoy going beyond the thousands groups to millions, but we seldom have need to read or write numbers greater than 999,999.

The writing of large numbers with words rather than numerals is seldom required, except for the writing of a personal check. Hence mildly handicapped children should practice that skill for numbers deemed reasonable by the teacher.

Rounding off Numbers

The value of estimation procedures has been increasingly recognized in recent years. Since machines now do much routine calculation, an emphasis on the ability to judge the suitability of the machine's output becomes a logical step. For the mildly handicapped to experience maximum benefit from such machines, skill in rounding off numbers is essential.

Skill in rounding off includes two components. The first of these is the ability to decide what level of accuracy is necessary. To say "It is about eighty miles from here to Cleveland" is acceptable. When giving directions to someplace six miles down the road, however, rounding distance to ten miles will probably make the directions useless. Likewise, in the preparation of income tax returns, rounding off to the nearest dollar is quite legitimate, but rounding off (especially rounding down) to the nearest one hundred dollars may well signal trouble for the future. Mildly handicapped children need many experiences in discussing situations that call for different levels of accuracy in rounding off.

The second component is the concept of *nearest to*. Understanding of this can be developed through the use of materials familiar to the child from the ongoing development of place-value concepts or through the use of number lines. When using a set of base-ten blocks, for example, the teacher can show the child a representation of 53 and ask if it is easier to remove the 3 units to show 50 or to add 7 units to show 60. On a number line, the teacher can mark 53 and ask the child to pretend he or she is standing at that point between two bases during a baseball game. All other factors being equal, toward which base would the child run—the base marked 50 or the base marked 60?

Experiences with both components can be provided formally and informally. Such activities should be made part of the instructional program early in the sequence and repeated frequently thoughout the sequence, as each new power of ten is introduced and developed. Through such steady exposure, mildly handicapped children will become fluent in their ability to round off numbers and will do so as a natural part of their experience with numbers.

Numbers as Labels

Ideally, we provide children with a wealth of experiences designed to help them understand numbers, with a view to better success with the algorithms of arithmetic. However, we use numbers very frequently in another, non-quantitative, context: the context of identification. Each of us has a unique Social Security number; drivers of cars and other vehicles have license numbers; towns and cities assign numbers to dwellings; pages in books have numbers; we number television channels; seats in an auditorium or at a sports arena have both letters and numbers; examples of numbers used as labels abound.

Mildly handicapped children need to be aware of such numbers and their use. A unit on identification numbers might include each child gathering the following information about himself or herself: Social Security number, house number, telephone number, parents' automobile registration number, and the like. A study of the numbers used by banks could form part of the study. A discussion of why numbers are more frequently used than letters for identification would also be profitable. Such a study would be incomplete without reference to computers and the relative ease with which they deal with numbers. All these experiences can serve to expand a mildly handicapped child's appreciation of the role of number in his or her world.

SUMMARY

The chapter began with reference to the many contexts in which we experience number. We ended with specific reference to the use of numbers as labels, as identifiers. In between we examined the young child's emerging understanding of number, rote, and the meaning of whole numbers, particularly as representing cardinal property. The skills necessary for meaningful counting were examined. We moved on to consider the system of notation we use and its reliance on a base-ten, place-value system. Different materials for developing a solid understanding of place value were presented. We then spent some time considering how to develop skill in rounding off numbers.

The development of a mature appreciation of number will require much instructional time and learning time for mildly handicapped children. The advantages of such knowledge are legion when viewed from the perspective of later mathematics content. It is to some of that content that we now turn our attention.

THE ARITHMETIC OF WHOLE NUMBERS

INTRODUCTION

Beyond a system of numeration, procedures for using numbers to tell an abbreviated story of what was happening in the marketplace and in one's own financial situation have been of interest to humans for many years. Menninger (1977, 351), with respect to the fifteenth and sixteenth centuries, wrote

> Many textbooks of the time list the following branches, or operations of arithmetic: numeration (the placing of the numbers on the counting board), *addition, subtraction, duplation* and *mediation* (that is, the doubling and halving of numbers), *multiplication,* and *division.*

Since 1700, two other operations have found their way into use: the process of raising numbers to powers (a specific use of multiplication) and finding roots of numbers.

Before we examine the operations of addition, subtraction, multiplication, and division, let us mention a distinction between the mathematical definitions of the operations and our customary ways of thinking about them. Each of the operations has a formal meaning, which, if not its own, is related to another operation. Thus, we hear subtraction spoken of as the inverse of addition, and division referred to as the inverse of multiplication. These formal definitions are good and necessary, but practical and interpretable descriptions of the operations and their uses are sufficient to most of us. Skill in translating these operations into meaningful processes for the children we teach is invaluable. This orientation characterizes the material in this chapter.

For each process we shall begin by examining the meaning of the operation. After reviewing the ways in which we normally use the operation under consideration, we will suggest a reasonable scope and sequence for mildly handicapped learners. Each section will conclude with suggestions for representing the operation manipulatively and pictorially, as well as explanations of different algorithms appropriate to each operation.

ADDITION

Addition is probably the easiest of the four operations to understand. Briefly, addition is a symbolic way to tell what happens when we join two sets that do not have any members in common. Addition is an operation we perform on numbers. The process provides us with a shorthand method for telling the story of what happens when we join sets. We do not add sets or things; we add only numbers.

Scope and Sequence

The decisions required for the determination of the scope or coverage of a topic should reflect, at a minimum, the use to which students will put the knowledge. Other considerations include the level of students' abilities, the breadth of their interests, and the demands future advances in technology may make. Hence we can provide only guidelines relative to the scope of any topic.

The average person in our society today uses addition for little more than balancing a checkbook and computing expenses. Therefore, with respect to addition, we suggest that mildly handicapped children, by the end of their academic career, should at least be able to add three addends of four or fewer digits when regrouping is required. Decisions to provide skills beyond that level should be made by those responsible for an individual student's mathematics program.

In Figure 6.1 we present a sequence reflective of research and the content of typical mathematics programs. Slight differences exist between sequences found in different programs; no one sacred sequence exists.

Most, if not all, commercially produced mathematics programs include a scope-and-sequence chart. Many times this chart is part of the teacher's guide. Many publishers also prepare scope-and-sequence information as part of their advertising materials.

For those special educators responsible for part or all of a student's mathematics program, familiarity with the texts used in the regular school program, including scope and sequence, will be beneficial. This information can serve not only as a guide to the student's instruction, but also as another link between special education personnel and the regular class teacher. This link offers good potential for improving both the quality of the youngster's program in the special education classroom and the transition back to the regular program when such a transition becomes desirable.

Additional sources of information about the sequencing of topics are achievement tests and diagnostic instruments. An examination of the skills items and concept items of a standardized achievement test will often reveal

1. Set union as represented by addition.
2. Adding two single-digit addends, sums less than 10.
3. Adding two single-digit addends, sums greater than or equal to 10.
4. Addition—role of zero.
5. Addition—three single-digit addends.
6. Cummutative property of addition.
7. Associative (grouping) property of addition.
8. Addition—two-digit numbers, no regrouping.
9. Addition—three-digit numbers, no regrouping.
10. Addition—two-digit numbers, regrouping.
11. Addition—three-digit numbers, regrouping from tens to hundreds.
12. Addition—three-digit numbers, two regroupings.
13. Addition—four-digit numbers, regrouping from hundreds to thousands.

FIGURE 6.1 Scope and sequence for addition

the task analysis that has gone into the selection of the items. Items on tests like the Brigance instruments can also serve the same purpose. Such information can be a valuable asset for the special educator, who may not have direct contact with regular education personnel and the tools of their trade.

Much of the foregoing can be applied to the desirable sequence for teaching each of the four operations under consideration. For the remaining three operations we will merely list appropriate sequences.

Approaches to the Teaching of Addition

Knowing how to add requires a set of concepts and a set of skills. Concepts are best taught by way of manipulatives; the algorithms of addition require an understanding as to the why of the rules and the how of putting the procedures on paper. To complicate the situation still further, certain skills are necessary in order to build concepts by manipulation, for manipulatives themselves often represent a level of abstraction.

The joining of sets affords an excellent behavior through which to effect the introduction to addition. Mildly handicapped learners should first have many experiences in combining the items from two or more sets of things. In the course of these activities the students should attend to the number of elements in each of the individual sets, and then the number of elements in the set that is the combination of the original sets. This larger number, then, can be defined as the sum of the cardinal properties of the two smaller sets. Through these experiences the child can be attaining familiarity with many of the addition facts, acquiring a means for representing the facts, or figuring out a combination that has not yet been committed to memory.

The learner should observe that the sum of two or more numbers is usually different from any of the addends. The exception to this rule occurs when one of only two addends is zero. By using representations of the empty set, the child can learn the role of zero in addition.

From using sets to represent addends and joining the sets to represent sum, the youngsters can move on to drawing pictures to illustrate the addends and counting all the items in the pictures to find the sum. From this point it is a small step to using tally marks to represent addends for unknown combinations and counting the tally marks to name the sum.

The addition combinations can also be represented on a number line. We recommend its use for two reasons. Many young children almost spontaneously use a number line of their own making to figure out the answers to unknown addition and subtraction combinations. A first-grade girl with whom one of the authors was working wrote the list of the numerals in a vertical column, starting at the top with 1 and going down to 20. When adding 8 and 7, she would locate 8 on her list, move down to 9, saying "one," and continue counting until she said "seven." At that time she was pointing to 15; she pronounced "fifteen" to be the answer. When asked who had taught the technique to her, she responded, "Nobody; I had to figure it out for myself." After observing her own seven-year-old son using a similar method for figuring out an answer to a subtraction problem, the author asked him how he had learned the method. His reply was, "I saw Jamie do it in school. It's easy." Resnick (1983, 110–111) writes:

> Several lines of evidence point to the probability that by the time they enter school most children have already constructed a representation of number that can be appropriately characterized as a mental number line. That is, numbers correspond to positions in a string, with the individual positions linked by a "successor" or "next" relationship and a directional marker on the string specifying that later positions on the string are larger. . . . This mental number line can be used both

FIGURE 6.2 Number Line

to establish quantities by the operations of counting and to directly compare quantities. By combining counting and comparison operations, a considerable amount of arithmetic problem solving can also be accomplished.

The second reason is provided by the need for some mildly handicapped students to move beyond whole numbers to all the integers in order to achieve even limited career goals. The number line supports the extension to negative numbers quite easily—even more so when the student is already familiar with its use. The number line is acquired easily and has great potential for representing fairly sophisticated concepts.

Number lines invented by youngsters do not necessary match the line illustrated at the top of Figure 6.2. Some children use only lists of numbers. Others may think of the numbers as blocks. These representations do not account for the presence of numbers between the whole numbers; the concept of the density of the real number system develops very gradually as new sets of numbers are introduced and used. The primitive number line of the young child can be modified as the need arises.

The use of manipulatives should not be confined to sets representing single-digit numbers. The addition of two-place numbers and three-place numbers should first find representation through the types of materials used for instruction in place value: the popsicle sticks, the base-ten blocks, the abacus, or the like. A good manipulative device can be used over a wide variety of concepts and over many levels. All of those mentioned have such potential.

THE ADDITION COMBINATIONS

The child who can use manipulatives to represent addition sentences and who can name an addition combination when it is represented is ready to memorize the combinations. Good instruction requires that the student be able to translate representations into symbols and vice versa. Meaningful drill and practice on the combinations can and should take many forms. Some of these are illustrated in the following examples. Each activity represents a different interaction between student and instructor.

1. Use a number of small items. These might be plastic chips, small straws, centimeter cubes, or the like. For each of the activities, begin by putting a few small items (four, for example) in one pile and another number of items (two, for example) in another pile. Next, combine the groups.

   a. Provide the children with a pool of small items. Ask them to do exactly what you did.
   b. Show the learner a display of sets that have different cardinal properties. After you join the two sets you have made, ask the learner to point to the set in the display that has the same number as your set.
   c. Ask the children to tell you the number of items in each of the smaller sets and the number of items in the set formed by joining the smaller sets.

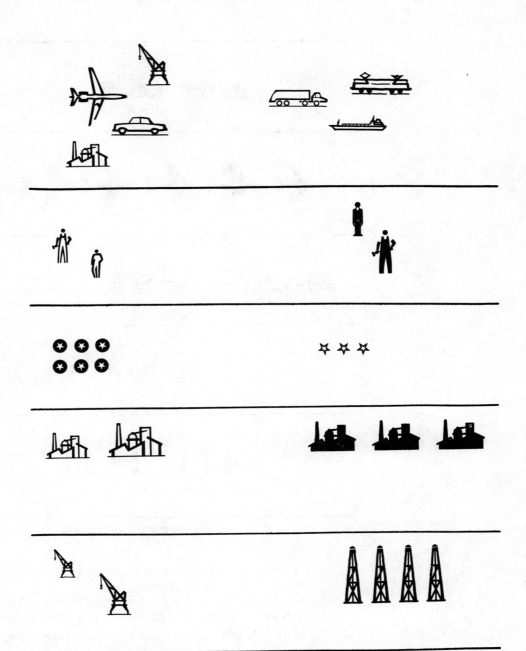

FIGURE 6.3 Sets to be joined

    d. Ask the learners to write the number of items in each smaller set and then the number of items in the larger set, or ask the learner to draw a picture of each smaller set and then to draw a picture of the set formed by the joining of the two smaller sets.

2. Display pairs of sets like those illustrated in Figure 6.3.

    a. Provide the children with a pool of small items. As you point to a pair of sets in the display, ask them to make a set showing how many altogether in the pair.

    b. Display sets like those shown in Figure 6.4. As you point to a pair of sets in the original display (Figure 6.3), the youngsters are to point to a set in the second display (Figure 6.4), showing how many items there would be in the union of the first two sets.

    c. Ask the children to tell you the number of members in each of the

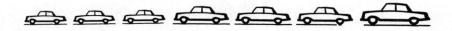

FIGURE 6.4 Joining sets

separate sets of each pair in Figure 6.3 and then to name the number of items there are in all.

   d. Ask the children to write the number of items in each of the sets in each pair in Figure 6.3 and then to write the number of items in all.

3. Dictate a pair of numbers—for example, "three, five."

   a. Ask the learners to make a set for each number dictated and then to join the two sets together.

   b. Present an illustration like Figure 6.3 and ask the children to point to the pair or set representing the two numbers dictated. Alternatively, present an illustration like Figure 6.4 and ask the children to point to the set representing the total of the two numbers dictated.

   c. Ask the children to say the number that tells how many in all for the two numbers dictated.

   d. Ask the children to draw sets representing the numbers dictated and to write the number telling how many in both sets together.

Students who successfully participate in activities like the preceding are ready to learn the notation we use when we write down the stories in shorthand. The numerals tell us how many in a set. To show that we are looking for the number of items in the union of sets, we use the plus sign, meaning "add." To separate the numbers being added from the sum, and to indicate that the sum represents the same quantity as the addition of the numbers, we use either the equals sign or the addition line. These symbols are useful because we do not want to have to make sets, draw pictures, or write words all the time when we want to figure out how many altogether in two sets or more. They are also worth knowing because we will use them in other contexts.

After the instructor provides the explanation and learners have had an opportunity to practice recognizing and writing the two symbols, the students are ready to participate in lessons like the following:

A C T I V I T Y

4. Distribute a worksheet on which examples of the basic addition combinations are written.
   a. Ask each child to make sets to represent each answer.
   b. Display several sets, each of which represents an answer to a different example. Ask the learners to match each example to the set representing its answers.
   c. Ask the learners to explain what each symbol in an example means.
   d. Ask the children to write the answer for each example or to draw a set representing the answer.

Memorizing the Addition Combinations

Mathematics educators generally agree that students need to learn the basic facts or combinations of addition, subtraction, multiplication, and division. Of the many reasons cited, two pertain to mildly handicapped children in particular. The learning of an algorithm for computation is greatly facilitated when the child can retrieve answers to the basic combinations quickly and accurately. Also, the use of a calculator for computation is enhanced when the user can evaluate a displayed answer for its appropriateness. Such assessment depends on being able to recall the basic facts easily. Hence a good mathematics program devotes much time and effort to enjoyable and meaningful drill and practice on the basic combinations.

Four qualities characterize successful drill and practice programs.

1. Opportunities for practice are regular, frequent in occurrence, and short in duration.
2. Practice exercises are enjoyable; undue stress is avoided.
3. Built-in, ongoing review marks the entire program.
4. Random guessing at answers is discouraged; children are encouraged to figure out the answer before responding to an item.

Encouraging learners to figure out answers before responding accomplishes two purposes. They avoid rehearsing incorrect responses, and they develop the conviction that they have the power to control the rightness or wrongness of an answer. That is, getting an item "right" is not solely a matter of luck. Many mildly handicapped learners are convinced that luck is the only component of good work.

Several means can afford children the opportunity to figure out answers ahead of time. Chips or other small objects can be given to students who would like to use them. Other children prefer to make tally marks on paper. The addition chart in Figure 6.5 is also useful. The chart serves to summarize all the basic addition combinations and to illustrate the effect the principle of commutativity has on the number of different combinations that have to be learned. Further, the skills needed to read such a chart are applicable to many different types of charts, including maps.

+	0	1	2	3	4	5	6	7	8	9
0	0	1	2	3	4	5	6	7	8	9
1	1	2	3	4	5	6	7	8	9	10
2	2	3	4	5	6	7	8	9	10	11
3	3	4	5	6	7	8	9	10	11	12
4	4	5	6	7	8	9	10	11	12	13
5	5	6	7	8	9	10	11	12	13	14
6	6	7	8	9	10	11	12	13	14	15
7	7	8	9	10	11	12	13	14	15	16
8	8	9	10	11	12	13	14	15	16	17
9	9	10	11	12	13	14	15	16	17	18

FIGURE 6.5   Addition chart

Many mildly handicapped children need to develop strategies for more efficiently arriving at goals. The computing of answers to a basic addition combination offers a context for instruction in this general ability. One strategy that can be used is the following. Suppose the item requiring a response is 5 + 8. The child can write the digit representing the larger quantity on a piece of paper, and to the right of that number write tally marks to represent the other number. By starting with the number written and continuing from there to count the tally marks, the child will arrive at the sum. The process is illustrated in Figure 6.6. Some children spontaneously use the same approach when finger counting. By representing the smaller addend with fingers and counting the fingers starting with one more than the larger number, the sum is attained more quickly.

A question that often arises concerns the order in which the basic combinations are to be mastered, and the number to be mastered at one time. The results of relevant research are not consistent, probably because the variables that affect the speed and accuracy with which a group of individuals respond to a set of combinations are many and nonobservable in most cases. Generally, combinations representing sums less than ten are mastered before larger sums. Doubles are easily mastered, and sums of numbers separated by one (for example, six plus seven) are mastered quickly if the child is taught to look for doubles and to add or subtract one.

Usually, the task of mastering all the combinations can be less daunting to a child if he or she can recognize that most of the facts are already known. If enough practice with manipulatives has been provided, the child should have in possession the combinations involving zero or one as addends. Already the first two rows and the first two columns of the addition chart can be eliminated. (Refer to Figure 6.5 again.) With a little practice, doubles are mastered. Once they are under control, all the sums named by the numbers next to each

$$
\begin{array}{ll}
8 & \text{``8''} \quad / \quad / \quad / \\
\underline{+3} & \qquad\quad \text{``9''} \ \text{``10''} \ \text{``11''}
\end{array}
$$

"The answer is eleven."

FIGURE 6.6   An efficient "counting to add" procedure

Chapter Six

number on the diagonal can be conquered. Now there are only twenty-one unique combinations remaining. They can be mastered directly, or combinations with ten as an addend can be introduced—for example, 10 + 4, 10 + 7, etc. These are easily learned, and the link to place value is clear. Once 7 + 10 is named as 17, 7 + 9 becomes 7 + 10 − 1, or 16. Now the number of combinations to be memorized is reduced to 15—hardly an insurmountable task.

Not every child may be able to do the types of thinking described above, but with appropriate experience with manipulatives, most students will not have to labor over fifty-five different combinations. The good instructor will be able to decide how much brute memorization is necessary and how many combinations can be reasoned out by an individual child. The point being made here is that the burden of the task can be lightened when the young person knows strategies for arriving at answers not immediately retrievable from his or her memory. Regardless of the approach taken, the goal should be swift retrieval of all combinations within a reasonable amount of instructional time.

Computer Assisted Practice

A frequently cited benefit of the use of computers in instruction is the patience of the machine and its capacity for endless repetition. Both of these traits suggest strongly the potential of computers for aiding the cause of drill and practice. Certainly many software products purport to provide children with expertise in arithmetic. In addition, there are many relatively inexpensive books containing programs, written in BASIC language, for developing arithmetic skills.

Programs differ in several aspects. The simplest serve only as a source of examples. The child is expected to provide the answer. In some cases the program will accept answers of two digits or more only when the answer is written from left to right. Most programs allow the child to "erase" an incorrect numeral, but some do not. Usually appropriate reinforcement follows a correct answer; one or more opportunities are provided to rework an example when the answer is wrong. Some programs keep a record of the child's progress and report that record to the child at the end of a session.

Of greater benefit are programs that combine practice on the basic number combinations with games requiring application of number concepts and strategy. Such a program is Number Stumper (Sorenson 1984). While providing practice on single-digit addition and subtraction facts, the game requires the player to name sums in several different ways and to select expressions that will result in the greatest number of points for the player as he or she eliminates possible choices for solution. Such games provide far more than drill and practice, and they reward the learner for original and/or divergent thinking.

In providing any software to children, instructors must be aware not only of the instructional goals and objectives of the product but also of any hindrances the game may present to successful performance. Some games require a lot of reading in order for the child to use the menu; others may refer to keys on the keyboard that do not precisely match those on the keyboard used by the child. With a little bit of help, most children can adapt, but caution is necessary if a child is not to meet yet another experience resulting in failure.

Calculators and Estimation

At this point two closely related issues deserve mention. The first is the growing use of hand-held calculators in our society and their benefits to mildly handicapped learners. The second item is the role of estimation skills in the day-to-day use of arithmetic. These two concerns are related, and their discussion finds a place in this present context of the memorization of number combinations.

In our conversations with teachers of mildly handicapped children, we have often heard the question "If a child has trouble learning the number facts, shouldn't we use instructional time for more important things and teach the student how to use a calculator?" The answer to the question depends on many factors. One of these is the age and educational experience of the learner. The mildly handicapped third grader is a different case from the secondary student with one or two years of formal academic training left. For the younger child we would encourage the use of every reasonable method to help in the memorization task. The successful use of the calculator requires that the user be able to recognize the answer obtained as being "in the ballpark" or not. That skill depends largely on the awareness of *about* where the answer should be, which requires knowledge of the basic number combinations.

What about the older student? Certainly we acknowledge the need for such a youngster to gain the maximum possible in the way of concepts and the practical uses of mathematics in his or her remaining years of school. For such a learner one is tempted to recommend training in the use of some mechanical means of computation. Here also, however, we would warn that the efficacy of such a tool depends to a large extent on the accuracy with which it is used and on the ability of the user to weigh the accuracy of the outcome. Training in the use of such tools must reflect these concerns.

Judging the accuracy of a computational result is one application of the skill of estimation. We all recognize the desirability, in many circumstances, of dealing in "ballpark figures" rather than in exact quantities. Estimation is a necessary and valuable skill. Few of us know exactly how much money we have in our checking account at any given moment, but most of us have an idea of *about* how much money there is. Few of us know exactly how much our purchases from the supermarket will cost us before we get to the cash register, but we know enough to accept or question the cashier's total. Exactly how much paint will be required to renovate the exterior of a house probably cannot be known before the completion of the job, but the careful painter will have a good estimate so that enough of the paint can be bought at one time and so that money will not be wasted on too much paint. We may not know exactly how much our order at a fast-food restaurant will cost us, but we know enough to judge the fit between the cash in our pocket and the total that will appear on the cash register.

Most of us have had the experience of waiting in line at a store behind a child whose purchases do not match his or her means. That same experience can be very embarrassing for an adult, or even an older student. Attention to the development of estimation skills can help an individual to avoid those situations. The prerequisite skills for estimation include the following.

1. Knowledge of the basic number combinations—at a minimum, those for addition, subtraction, multiplication.
2. The recognition that for estimation purposes, those numbers representing larger powers of ten are more important than the digits representing smaller powers of ten.
3. Skill in reading and computing with numbers (at least to a limited degree) from left to right.
4. Skill in recognizing the degree of accuracy appropriate in a given situation.

A good mathematics program will attend to the development of these skills over the entire course of a student's academic experience.

Manipulatives and Algorithms

We are about ready to examine some of the algorithms (rules for computation) for addition. Before we do, however, we would like to point out some discrepancies between what we do with manipulatives and the order in which we write in shorthand the "story of what we do." For purposes of illustration, we ask the reader to use some materials (base-ten blocks, abacus, chips) to find the answer to this addition example: 436 + 122 = _____ .

For the sake of simplicity we shall discuss the example in terms of base-ten blocks. There are several ways in which you could have represented the example. You could represent 436 first, followed by the placement of the representation of 122 under it. After the addends are represented, you can show the sum by moving the hundreds-flats together, then the tens, then the ones, or you could move first the ones together, then the tens, then the hundreds. Needless to say, the tens could be joined first also. Note also that the addends can be represented by showing hundreds first, tens first, or ones first. Experiment with other ways of representing the addition.

Just as the number of ways in which we can represent examples manipulatively are many, so the number of ways in which we can carry out the symbolic computation are also many. Some are perhaps more efficient than others; some may occupy less space on a sheet of paper than others; some may be more in tune with our level of comfort and understanding than others. None, however, is the only correct way.

Also important is an examination of the match between the different forms of manipulative representation and the format in which the example would be written. Many mildly handicapped children can make the transition from manipulatives to symbols even when the order of placement or events in one mode does not match that in the other mode. Other mildly handicapped learners, however, need explicit instruction on the differences between the steps that might be used during manipulation and those that might take place during computation.

Algorithms: Rules for Computation

Ideally, the procedures used in computation should be arrived at by the child after many experiences in finding sums with materials. The exercises should include examples with up to three addends of three digits each. Such a sequence permits the child to invent and modify his or her own rules until a satisfactory method is reached. With proper guidance each child will develop an algorithm that is both effective and efficient.

Are mildly handicapped children capable of such invention? There is ample evidence that many students, including mildly handicapped children, invent algorithms even when taught the traditional algorithms. Although the invented algorithms are often faulty, most of the invented procedures follow rules. The inaccuracy of the procedures usually results from a lack of understanding of the symbols being used. Hence we have every reason to suspect that given proper exposure and guidance, children will create their own effective and efficient algorithms.

If a proper understanding of place value has been developed, and the child has had sufficient opportunity to find sums by manipulation, it is very likely that the algorithm he or she develops will resemble one of those summarized in Figure 6.7. The top algorithm is the traditional one. Addition is

A. *Traditional algorithm*

Example	①	②	③
	1	1	1
412	412	412	412
236	236	236	236
+247	+247	+247	+247
	5	95	895

B. *Left-to-right algorithm*

Example	①	②	③
412	412	412	412
236	236	236	236
+247	+247	+247	+247
	8	88	885
			1
			895

C. *Expanded notation*

Example	①	②
412	$400 + 10 + 2$	$400 + 10 + 2$
236	$200 + 30 + 6$	$200 + 30 + 6$
+247	$+200 + 40 + 7$	$200 + 80 + 7$
		$800 + 80 + 15$

③
$$800 + 80 + 15 = 800 + 90 + 5 = 895$$

FIGURE 6.7 Different addition algorithms

accomplished from right to left, and regrouping is done as the sum of each column is completed. The second algorithm illustrates the process of adding from left to right. Regrouping is taken care of by adding the multiple of the higher power of ten to the column previously added. The third algorithm represents the approach taken when the child writes each addend in expanded form. As indicated, the child can add from right to left or from left to right, regrouping as appropriate.

Clearly, the traditional algorithm is most efficient from the standpoints of space needed to write down the story and amount of ink used. For many centuries of human computing those two considerations were very important. Paper as we know it today and ink (or other recording media) were very scarce. For those reasons algorithms which served to lessen the amount of writing that had to be done received more attention than those that did not.

Present-day technology has made such considerations superfluous. With calculators to compute for us, it is essential that we honor teaching strategies that develop both fluency with *an* algorithm and a basic understanding of what computation is all about. In many senses the special education teacher is in an enviable position to adopt such an attitude with his or her students.

Allowing the mildly handicapped child to "invent" an algorithm and to test his or her procedures for efficacy provides added benefits. Through such a process the child should come to recognize that mathematics is not an arbitrary system of rules, but a meaningful system which admits of many variations. Quite often, when they are used, movable materials are employed to justify the traditional algorithm. While such efforts are superior to teaching the algorithm as a sequence in which to manipulate symbols only, such an attempt may serve only to impress upon the child the arbitrary nature of mathematics. For the mildly handicapped child, that can lead to fear and frustration, a result no teacher wants to cause. Much of mathematics has been "created" by humans as different needs in dealing with our quantitative process arose. Children, even mildly handicapped youngsters, must be able to participate in that creative endeavor.

SUBTRACTION

Most teachers of elementary school children will agree that children seem to find subtraction harder than addition or multiplication. Testing designed to measure the progress children make in learning different skills confirms this observation (NAEP 1974). One reason for the difficulty may be found in the multiplicity of situations in which we use subtraction to arrive at an answer.

We use the operation of subtraction to find out what is left in a set after some members have been removed. Typical events of this type would be these:

- A teacher had 27 students in a classroom. Seven students left the classroom to go to a special music class. How many students were left in the room?
- Mary earns $1,200 a month. How much money does she have left after she pays her rent which is $450?
- A painter bought 4 gallons of paint. He used 2 gallons on a garage. How much paint remained?

We also use subtraction to find the magnitude by which the number of elements in one set differs from the number of elements in another set. For example:

- There are 32 students in the fourth-grade class and 27 students in the fifth-grade class. Which class is the larger? By how much?
- Marsha earns $950 a month. Her sister earns $1125 a month. How much larger is Mary's sister's salary?
- The painter needed 7 quarts of paint for the back porch of a house and 4 quarts of paint for the front porch. How many fewer quarts did the painter need for the front porch?

Finally, we use subtraction to determine the magnitude of what has to be joined to an existing set to meet some demand for a larger set. Typical situations falling into this category would be these:

- A teacher's class was raising money to go on a class trip. They needed to earn $2,700. So far they had raised $1500. How much money do they still need?
- Mary needs a raise. She earns $1350 a month but her expenses come to $1500 a month. How large a raise does Mary need?
- A painter bought 4 gallons of paint for a house that required 9 gallons. How much additional paint did the painter have to buy in order to complete the house?

The three groups of situations listed above summarize and clarify the different types of thinking children must bring to using subtraction in an applied way. From the perspective of representing subtraction expressions by manipulation, the first type mentioned, sometimes called *remainder subtraction*, is the easiest to demonstrate. Unfortunately, many of us stop there. Mildly handicapped children will benefit from representing all three types.

A reasonable scope and sequence for teaching subtraction is summarized

1. Subtraction represented by the removal of a proper subset of a given set.
2. Subtraction—top number less than ten.
3. Subtraction—top number equal to or greater than ten but less than 19.
4. Subtraction—not commutative.
5. Role of zero in subtraction.
6. Subtraction—multiples of ten (for example, $70 - 40 = ?$).
7. Subtraction—multiples of one hundred.
8. Subtraction—two digit numbers, no renaming.
9. Subtraction—three-digit numbers, no renaming.
10. Subtraction—four-digit numbers, no renaming.
11. Subtraction—two-digit numbers, renaming one ten as ten ones.
12. Subtraction—three-digit numbers, renaming one hundred as ten tens.
13. Subtraction—three-digit numbers, two renamings.
14. Subtraction—three-digit numbers, 0 in ones place of top number.
15. Subtraction—three-digit numbers, 0 in tens place of top number.
16. Subtraction—top number a multiple of 100.
17. Subtraction—four-digit numbers, renaming one thousand to ten hundreds.
18. Subtraction—four-digit numbers, three renamings.

FIGURE 6.8   Scope and sequence for subtraction

in Figure 6.8. As was the case with addition, the skills listed do not necessarily imply the acquisition of the traditional algorithm.

Teaching Subtraction

Because the different applications of subtraction lend themselves to manipulative representations differently, we shall discuss each separately.

Remainder subtraction.   Since this subtraction situation implies the existence of a beginning set, representations of items in a set form a natural first step. The number of elements in the set is written down as the top number (or the left number) in the example. The number of items to be removed from the original set is represented by the bottom number, and the number of items remaining is the answer.

The terms *top number, bottom number, left number,* and *answer,* while not elegant, are quite appropriate during instruction. Some programs use the terms *sum, known addend,* and *missing addend,* respectively. Because these terms represent only the symbols themselves, regardless of application, we prefer the simpler terms during initial instruction.

For the basic combinations, the top number in an example can be represented by small items: paper clips, stars, chips, or whatever is at hand. Larger numbers can be represented by chips of different sizes and/or colors, with base-ten blocks, popsicle sticks, or on an abacus. For the basic combinations or for examples in which the number to be subtracted is a one-digit number, the number line offers an efficient method.

To use the number line, the top number is located first, or placed first. The bottom number can be represented by taking steps "backward" on the number. The number located at the end of the last step is the answer. Note that the steps are counted, not the points on the line. An alternate method, of course, is to start with the bottom number of the example and count the number of steps necessary to move forward to the top number. That number of steps is the answer.

Comparison subtraction. Clearly, for small numbers, the representation of the two sets to be compared is the preferred way to illustrate this use of subtraction. The members of one set are to be matched with the members of the other set until no more pairs can be made. By removing the pairs and counting the items that remain, we can determine the difference between the two sets.

Demonstrating the comparing of two larger numbers through the use of base-ten blocks is effective as long as the comparison requires no renaming. Units can be paired with units, tens with tens, and so on. Once again the answer will be represented by the blocks that "have no partners" after the pairing is completed.

This kind of exercise serves only to justify the use of subtraction to determine the quantitative difference between two sets. Essentially, we remove or take away the members of the larger set that can be paired with members of the smaller. The remainder tells us how many more there were in the larger set.

Additive subtraction. The representation of the third application requires the awareness of the hypothetical larger set. Some students do this naturally in their finger counting for subtraction. For example, when asked to find the answer to an expression such as $9 - 6$, some learners will hold up six fingers and then continue counting from 6, holding up the necessary fingers, until 9 is reached. They derive the answer by noting how many extra fingers were needed. They seem to be reasoning $6 + ? = 9$.

When larger numbers are necessary, other means of visualizing the set having the desired cardinality must be used. For such representation the number line can be effective. The learner can locate the number symbolizing the cardinality of the desired set and then the number representing the cardinality of the set that exists. By counting the number of steps from one number to the other the student can arrive at the answer. Two number lines, one under the other, can help to relate the number line representation to the set representation. Essentially, the answer is found by counting how many steps are left on the top number line after removing what matches the bottom number line. Experiences such as those described may help a child to understand why subtraction "works" in the context under consideration.

A C T I V I T Y

TEACHING THE SUBTRACTION COMBINATIONS

The reader will note that many of the activities suggested for teaching the addition combinations are useful for teaching the subtraction combinations. We will mention here only activities that for one reason or another are particularly useful for drill and practice in subtraction.

1. The addition chart can be used for finding the answer to subtraction expressions, since they are derived from the addition combinations. The child locates the bottom number in the leftmost column, and then moves his or her finger along that row until he or she reaches the top number in the example. By moving a finger up that column to the very top, the child will locate the answer.
2. Write a list of several incomplete subtraction sentences, all of which have the same answer; for example, $15 - 9 = $, $14 - 8 = $, $13 - 7 = $,

etc. Direct the children to complete the list by adding other incomplete sentences having the same answer. When the list is complete, or at another time, give the children only the answer and the highest number that can be used in any example (assuming only whole numbers are to be used). For the example given, the answer would be named as 6 and the highest number to begin with is 15. Require the children to write all the subtraction sentences that meet the conditions given. They should respond with all the sentences in the list already developed.

3. Prepare a list of three numbers. The numbers in each group should be such that they can be rearranged to make a complete subtraction sentence. The learners' task is to write two subtraction sentences for each group of three numbers. For example, for the set 2, 4, 6, the children should respond with $6 - 2 = 4$ and $6 - 4 = 2$.

4. Show the children a subtraction sentence. Ask them to make up three stories, one using the sentence in a remainder situation, one using the sentence in a comparison situation, and one using the sentence in an additive situation.

Subtraction Algorithms

As was the case for addition, unless a child wishes to do some record keeping on his or her own, children should have a great deal of experience representing subtraction examples with manipulatives before the need to write anything down is imposed. To build correct habits from the beginning, the children, when given a subtraction example, should be encouraged to represent the top number first and to remove from that representation a representation of the bottom number. At the same time children should be reminded that at any point in their work, a ten-rod (assuming the use of base-ten blocks) can be exchanged for ten units, a hundred-flat can be exchanged for ten ten-rods, and so on. This reminder should be repeated when children first encounter examples for which renaming is required.

Again, assuming solid understanding of place value and the meaning of subtraction, children will develop valid sequences of steps to find answers manipulatively. In some cases these sequences are easily translated into the traditional right-to-left algorithm; in others the translation will effect a less traditional left-to-right procedure. Either direction is acceptable, as long as the child understands what the symbolic representation means because he or she has "invented" it. We summarize the two algorithms already mentioned, as well as one other nontraditional algorithm, in Figure 6.9.

MULTIPLICATION

We sometimes think of multiplication as a shortcut to finding the sum of the cardinal properties of several equivalent sets. Thus, if every seat in every row of a theater were occupied, and if each row had the same number of seats, we would multiply the number of rows by the number of seats in each row to find the number of people in the theater. In such a case, one factor represents the number of sets, and the other factor represents the number of members in each set.

However, there are other classes of situations to be examined:

• How many different sundaes can be made using only three different kinds of ice cream and four different toppings, assuming only one kind of ice cream and one kind of topping can be used on each sundae?

Example
932
−753

Decomposition algorithm

Step 1
```
    2
9 3̷12
−7 5 3
        9
```
Rename 1 ten as 10 ones and add them to the 2 ones. Subtract 3 ones from 12 ones.

Step 2
```
   812
9 3̷12
− 7 5 3
      7 9
```
Rename 1 hundred as 10 tens and add them to the 2 tens. Subtract 5 tens from 12 tens.

Step 3
```
   812
9̷ 3̷ 2
− 7 5 3
  1 7 9
```
Subtract 7 hundreds from 8 hundreds.

Left-to-right algorithm

Step 1
```
9 3 2
−7 5 3
 2
```
Subtract 7 hundreds from 9 hundreds.

Step 2
```
913 2
− 7 5 3
    2̷ 8
  1
```
Rename 1 of the hundreds in the answer as 10 tens and add them to the 3 tens. Subtract 5 tens from 13 tens.

Step 3
```
91312
− 7 5 3
  2̷ 8̷ 9
  1 7
```
Rename 1 of the tens in the answer as 10 ones and add them to the 2 ones. Subtract 3 ones from 12 ones.

Answer: 179

Equal additions algorithm (based on the principle that if the same number is added to both terms of a subtraction expression, the result is the same as that obtained from the original expression; for example, 27 − 5 = 22; 27 + 3 − (5 + 3) = 22).

Step 1
```
9 3̷12
−76̷5 3
       9
```
Add 10 ones to 2 ones. Add 1 ten to the 5 tens. Subtract 3 ones from 12 ones.

Step 2
```
91312
−8̷76̷5 3
      7 9
```
Add 10 tens to 3 tens. Add 1 hundred to 7 hundreds. Subtract 6 tens from 13 tens.

Step 3
```
91312
−8̷76̷5 3
  1 7 9
```
Subtract 8 hundreds from 9 hundreds.

FIGURE 6.9   Subtraction algorithms

- How many different outfits can be made from five different tops and three different skirts, assuming that each top can be worn with each skirt?
- How many different intersections will be made by seven avenues running east-west and six streets running north-south, assuming each avenue intersects every street?

Each of the solutions is illustrated in Figure 6.10. Symbolically, each answer is also arrived at through multiplication. The latter interpretation is more closely akin to the mathematical definition of multiplication.

Mildly handicapped children will benefit from experiences designed to promote understanding of the use of multiplication in both situations. Because the former interpretation of multiplication is used more frequently, we will discuss the representation of multiplication from that vantage point. We shall then make brief mention of the mix-and-match representation.

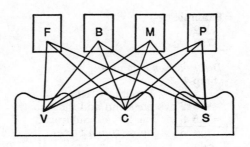

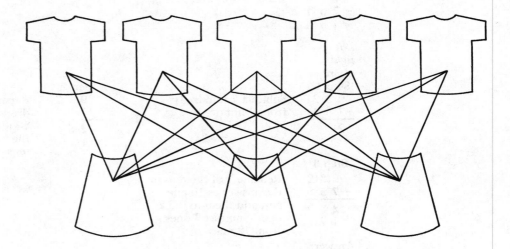

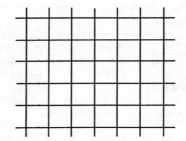

FIGURE 6.10   Model for multiplication

Representing Multiplication

Multiplication can be represented through manipulatives or pictures in several ways. Some of the more widely used methods are shown in Figure 6.11. In each case the expression 3 × 5 is represented.

In each case involving equivalent sets or arrays, the order of the factors is reflected in the diagram through a general rule which states that the first factor names the number of sets or rows; the second factor names the number in each set or row. While not important from a pure mathematical standpoint, the distinction serves to provide a means of translating word problems into multiplication expressions. A second advantage is found in the ease with which the same interpretation applies to to multiplication expressions involving fraction or decimal factors. That is, just as 3 × 4 can be read as three groups of four items each, $\frac{1}{3} × \frac{1}{4}$ can be translated into one-third of a one-fourth portion. For those mildly handicapped children who will attempt basic

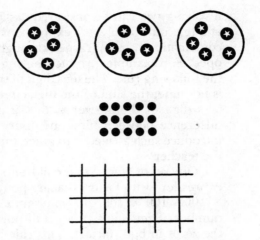

FIGURE 6.11   Representations of 3 × 5

algebra at some point in their academic career, a third benefit to be derived from careful translation of the notation is its application to terms such as 3*y*. Youngsters who are accustomed to thinking of the role each factor plays will find it easier to think of the symbol 3*y* as 3 groups of *y* elements each.

Of special importance when introducing multiplication is the inclusion of expressions having zero as a factor and the use of some representation of the empty set. Children can be led to see that any number of empty wallets leaves us with no money; any number of empty gum wrappers leaves us with no gum, and so on. Hence whenever we have zero as a factor, the product is zero.

The use of intersecting strips conveys visually the notion of multiplication as a cross-product. The intersections represent the "crossing" of each item in one set with the items in another set. This form of illustration is especially adaptable to picturing the traditional multiplication algorithm. It can also be used to represent products such as $\frac{1}{4} \times \frac{1}{3}$, as we shall observe in a later chapter.

The Multiplication Combinations

All the suggestions given previously for addition and subtraction apply to instruction/learning of the multiplication facts. The multiplication chart shown in Figure 6.12 is like the addition chart in format. After many experiences in representing single-digit multiplication expressions, children should be ready to commit the answers to memory.

As was the case with addition, the task of memorizing the multiplication facts will be much less daunting if youngsters recognize the large number of facts they have already acquired. The combinations involving zero and/or one provide no trouble; if children have learned to count by twos, the expressions

×	1	2	3	4	5	6	7	8	9
1	1	2	3	4	5	6	7	8	9
2	2	4	6	8	10	12	14	16	18
3	3	6	9	12	15	18	21	24	27
4	4	8	12	16	20	24	28	32	36
5	5	10	15	20	25	30	35	40	45
6	6	12	18	24	30	36	42	48	54
7	7	14	21	28	35	42	49	56	63
8	8	16	24	32	40	48	56	64	72
9	9	18	27	36	45	54	63	72	81

FIGURE 6.12   Multiplication chart

The Arithmetic of Whole Numbers

involving two can be answered easily. Most children learn to count by fives and tens quite easily; the expressions with five or ten as a factor follow. Other combinations can be learned through practice, or helping rules can be developed. For example, products having nine as a factor can be arrived at using the following rule. The tens digit of the product is one less than the factor that is not nine; the sum of the digits in the product must equal nine. Hence the tens digit of the answer to 5×9 must be 4. The ones digit must be the difference between nine and four, or 5. The decision whether or not to introduce such a rule, or to guide the children to discover the rule, rests with the teacher.

One set of "facts" that children must acquire for multiplication are those expressions with factors that are powers of ten: 10×1, 10×10, 10×100, 10×1000, 100×100, at a minimum. The basic rule is easily discovered: the number of zeros in a product of powers of ten can be arrived at by using all the zeros in both factors. This rule is necessary both for understanding the traditional multiplication algorithm and for skill in estimating products.

Multiplication Algorithms

The expression 4×217 can be represented using base-ten blocks. The product can be derived from making any necessary exchanges (for example, exchanging 20 units for 2 ten rods) and then counting the representations of the different powers of ten. The process, however, differs little from representing $217 + 217 + 217 + 217$. How do we move from here to building a basis for the conventional algorithm?

The conventional algorithm for multiplication is not easily represented by way of manipulation, although the meaning of an expression such as 36×712 can be interpreted as 36 groups of 712 items. The conventional algorithm can be interpreted as the symbolic cross-product of the multiples of powers of ten in one factor with the multiples of powers of ten in the other factor. In Figure 6.13 we have illustrated such a cross-product, using strips of different widths to represent different powers of ten; we have also attempted to show the parallel symbolic computation. In such a way the rationale behind the traditional algorithm can be developed.

Will a child "invent" the traditional algorithm as a result of working out examples using the crossing of strips, as illustrated? The conventional algorithm exemplifies the point made earlier in the chapter about the adoption of those methods of computation that required the least amount of recording.

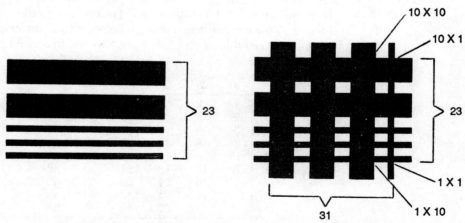

FIGURE 6.13 Representation of a multiplication algorithm

Chapter Six

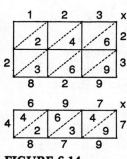

FIGURE 6.14
Lattice multiplication

Through manipulative representation the child will gain an understanding of the role that powers of ten assume in the algorithm and of the need for partial products. Also the child can record his or her results from the manipulation. After those skills have been achieved, the conventional algorithm can be introduced as the shorthand method it really is, if facility with the traditional algorithm is desirable for the child.

A somewhat more mechanical algorithm that does much to aid children organize the placement of the digits is the diagonal lattice procedure. This method eliminates the need for regrouping while multiplying. Outlined in Figure 6-14, the lattice calls for the placement of each digit of each partial product in a portion of a box. The child then adds the digits between the pairs of diagonal lines to find the total product.

The placement of the digit above the diagonal when necessary takes care of the "carrying." The algorithm is characterized by a different format for writing the story of what happens when we multiply, rather than a gimmick. Some mildly handicapped students find the structure provided by the boxes helpful; others find the format confusing. The introduction of the algorithm would reflect the instructor's opinion that in a given situation, the algorithm would be of benefit to a particular child.

DIVISION

The two major types of situations calling for the application of division are easily explained; the algorithm usually used to calculate the answer to an expression is difficult to learn for most children. Since the need for "long division" in our everyday affairs is somewhat limited, our discussion of division will focus on the meaning of division; we will only briefly outline a scope and sequence, one based on a repeated subtraction.

Reflect on the following situation. A decorator has 24 yards of drapery material and an order for pairs of drapes, each of which requires 4 yards of material. How many pairs of drapes can the decorator make with the material at hand?

Imagine yourself role-playing the solution to the problem. You would probably lay out the 24 yards of cloth and cut it into pieces of 4 yards each. You would then count the number of pieces you get.

If we translate the problem into a number sentence, we would get the following: n pieces of 4 yards each is 24 yards, or $n \times 4 = 24$. Traditionally we find the value of n by dividing 24 by the other factor, 4. This type of division, one in which we know the size of each group but not the number of groups, is called *measurement division*.

Let us return to our role-playing. This time the decorator has the same 24 yards of cloth on hand, but the order is for 8 cushion covers. How many yards can be used for each cusion cover if the decorator wants to use all the fabric?

To solve this problem without computation, we might assign one yard of cloth to each cushion and then repeat the process until all the cloth has been assigned to the cushions. In practice, we would probably use some representation of the cloth, to avoid cutting the cloth into useless pieces.

Translating the problem into a number sentence would result in this: $8 \times n = 24$. This time we know the number of groups or pieces we need. We do not know how large each piece or group would be. The term used for this type of division is *partitive division*.

As illustrated in Figure 6.15, we can use equivalent sets, arrays, and the number line to illustrate simple division examples. Instead of counting to determine the total number of elements involved, we count to find the number of groups or the number in each group. Thus, to represent 12 divided by

The Arithmetic of Whole Numbers   **113**

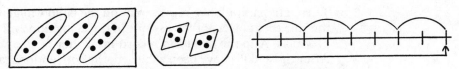

FIGURE 6.15 Representation of division

3, we might count out sets of 3 until all 12 items have been used, or we can "deal" 1 item to each of 3 sets until all 12 items have been used. In the former case we would count the groups to get the answer; in the latter we would count the number in each group after all the items have been dealt. When working with arrays, the groups become rows in the array. On a number line the groups become the jumps that are taken.

The following activities are designed to help the child to become familiar with the meaning of division, and to aid the development of a means of solution to simple division problems. At the same time the student can be learning the basic division combinations.

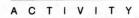

1. Place a set of small items on a table—for example, 12 plastic chips. Around each subset of 2 items, place pieces of string arranged to enclose the subsets.

   a. Provide the children with pools of small items and pieces of string. Ask them to do what you did.

   b. Display representations of different sets divided into subsets, one of which matches the division you have represented. After you arrange the subsets as described, ask the children to point to the matching representation on display.

   c. After you represent each division situation, ask the learners to describe aloud what you did. Encourage the children to name the number of items in the original set, the number of subsets, and the number of items in each subset.

   d. Ask the children to write the number of items in each subset, or to draw a picture showing what you did.

2. Display arrangements like those in Figure 6.16. For the activities, point to the pictures at the top of the figure, one at a time.

   a. Provide the children with appropriate materials. Direct them to make copies of each representation as you point to it.

   b. As you point to each representation at the top of the display, ask the children to find a matching representation at the bottom of the display.

   c. As you point to each representation, ask the children to describe it. Encourage them to name the number of items in the entire representation, the number of subsets, and the number of items in each subset.

   d. Ask the children to write the number of items in each subset, or to write the number of subsets.

3. Describe to the children situations to which division can be applied. How many cards would each of 4 people get if there were 24 cards to begin with? How many teams of 5 people each could you make if you had 15 people willing to play the game? How many gumballs costing 5 cents each could you get from a machine if you had 35 cents, all in nickels?

   a. Provide the children with suitable materials. Ask them to demonstrate how they would solve each problem you dictate.

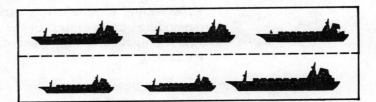

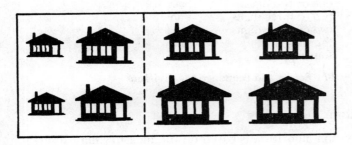

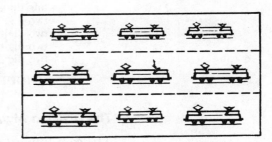

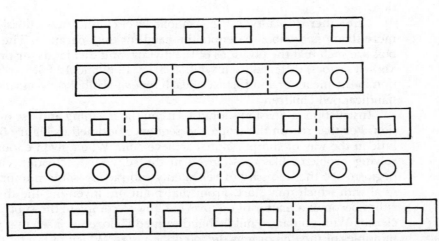

FIGURE 6.16 Representations of division expressions

b. Display representations of division expressions, one for each example you describe. Ask the learners to point to the appropriate representation each time.

c. Ask the children to describe in words the ways in which they might solve each problem.

d. Ask the children to work together as a group to draw a picture showing how each problem might be solved. As the children become proficient in this task, suggest that they use numbers to represent what they are doing.

4. Provide the children with written stories like those used previously. Repeat the directions given for activities 3a–3d.

1. Representing "measurement" division—for example, showing $28 \div 7$ by removing a group of 7 from a group of 28 and repeating the action three more times.
2. Representing "partitioning" division—for example, showing $28 \div 7$ by dealing out 28 cards to 7 individuals, one card at a time.
3. Representing remainders as part of another set or as single items, depending on the context of the situation.
4. Performing the following types of division examples by an appropriate method:

Divisor	Dividend	Remainder?
one digit	one or two digits	no
one digit	one or two digits	yes
two digits	two or three digits	no
two digits	two or three digits	yes
two digits	four digits	no
two digits	four digits	yes

FIGURE 6.17 Scope and sequence for division

The Division Algorithm

The conventional division algorithm is based on the measurement interpretation of division. Long division is basically repeated subtractions of multiples of the divisor. The multiples are obscured by the format of the calculation.

One other difficulty at the symbolic level characterizes division. We use more than one symbol to denote the need for the operation. The three symbols are such that the positions of the divisors and dividends (or products and known factors) differ in each symbol: $3\overline{)12}$, 12/3, and $12 \div 3$. Explicit instruction in the meaning and format of each type of symbol is necessary for mildly handicapped children.

Instruction in division should focus on the meaning and use of the operation. For that reason the scope and sequence outlined in Figure 6.17 reflects little in the way of computational achievement. When given enough practice solving problems like those given in the suggested activities, children will begin to see the necessity of repeatedly subtracting some number from the total with which they start. From that point on, inventing the division algorithm becomes a matter of discovering how to complete the subtractions efficiently. With guidance, mildly handicapped children will see that subtracting multiples of the divisor gets the job done faster. Which multiples they choose to subtract does not really matter, unless they want to proceed in the fastest manner possible. Since a calculator is faster than we are when computing answers to lengthy examples, emphasizing the procedure rather than the abbreviated algorithm seems to be a more valid approach.

SUMMARY

With machines to do basic calculations, the role of humans in the arithmetic process of problem solving becomes one of understanding the procedure and evaluating the results. This basic principle has guided us in our approach to the teaching of the arithmetic of whole numbers to mildly handicapped children. Instruction should concentrate on the development of understanding of the numbers and operations, the acknowledgment of a child's freedom to invent necessary algorithms, and the provision of experiences through which that freedom can be exercised. The creation of such an environment, the knowledge of the basic number facts, and the skills necessary to evaluate results properly, are the major objectives for instruction. This same philosophy will also guide our approach to the material in the next chapter.

FRACTIONS

INTRODUCTION

Most of us think of fractions as symbols used to name parts of a whole. The part–whole relationship, then, is an insight basic to working with fractions. In our work with mildly handicapped young children, we have found that formal and informal instruction in parts and wholes enhances later instruction in fractions and decimals. For this reason we begin this chapter with a discussion of the part-whole relationship and suggestions for activities to develop this concept with children. We then progress to an examination of the traditional content in fractions and decimals. We conclude the chapter with a brief look at suggestions for teaching the basic concepts involved in percent to mildly handicapped children.

PART-WHOLE RELATIONSHIPS

Perhaps the easiest part-whole relationship for children to grasp is the relationship between the discrete parts of the whole and the whole itself, which is composed of those parts. Several puzzles suitable for the preschool child typify this relationship. Young children learn to recognize ears, the head, the body, the tail, and the four legs as comprising the whole called dog. The head, neck, torso, arms, hands, legs, and feet are placed properly to form the human figure. Children of preschool age also learn to identify missing discrete parts of a whole.

A secondary benefit resulting from these experiences is the awareness that some parts are wholes when looked at in and of themselves. Thus, the hand is a whole that is comprised of fingers, a thumb, and a palm. This concept is important and finds application in later, more formal work with fractions.

Parts of wholes are not always separate from each other. We employ the term *blended parts* to refer to such mixtures. Blending can have two results. Sometimes the parts can be separated from each other after the mixing has occurred, and sometimes the components cannot be readily recognized or separated. A mixture of salt and pepper or of paper clips and pebbles in a jar falls into the first category. A cake mix falls into the latter group.

DEVELOPING AWARENESS OF THE DIFFERENT KINDS OF PART-WHOLE RELATIONSHIPS

1. Assemble part of a puzzle composed of discrete parts of a whole. Omit one or two pieces. Display puzzle parts, some of which are the parts missing from your puzzle. Direct the children to select and to place the piece(s) necessary to complete your puzzle. When the children become familiar with the task, assign one child to assume the role of the instructor. Select another child to complete the puzzle. If correct, the second child becomes the leader.

2. Display several puzzle parts, only some of which can be assembled to form a whole. The children are to identify those pieces and put them together to form the whole.

3. Outline a familiar figure on a sheet of paper or on a chalkboard. Ask a child to complete the picture by drawing the parts of the figure.

4. Remind the learners that a part of one whole can be considered as another whole having its own parts. Assemble a puzzle of a door; that is, cut from paper or cardboard a picture of a door, a doorknob, and a doorknocker. Talk about the wholeness of the figure called door. Then make the door a part of a bigger figure—a house. Suggest that the children draw a picture of a house and that they use a dark crayon or pen to outline all the parts of the house. After all the parts have been outlined, the children can color the parts of the house.

5. Display a picture of an incomplete puzzle that has the missing parts outlined. Give the children a set of all the pieces that could be used to construct a copy of the puzzle in the picture. Ask the children to find the pieces that are missing and to fit them into the appropriate places on the picture. Discuss with the children the fact that only certain parts can complete the whole.

6. To exhibit different ways in which materials can be mixed, obtain two containers of water, some sugar or salt, and some paper clips or safety pins. Add the sugar or salt to one container of water and stir the contents. Ask a learner to add the paper clips or safety pins to the other container of water and to stir these contents. Discuss the results. Why can't the sugar or salt be seen any longer? What happened to it? Introduce the word *dissolve* if necessary. What other things might also dissolve in water? Why didn't the paper clips or safety pins dissolve? Would they dissolve if left in the water for a longer length of time? Would the mixture of water and sugar or water and salt taste the same as the water mixed with paper clips or safety pins? Why?

7. Suggested materials: four clear one-cup containers; for example, two glass measuring cups and two clear pint containers. Mix food coloring and one cup of water to make a blue water liquid. Mix food coloring and another cup of water to make a yellow liquid. Mix the two cups of colored liquid in a pint container to make a green liquid. Demonstrate that when the pint is emptied into the cups, each small container is filled. In this case one cup plus one cup does equal two cups. Repeat the activity, but this time fill one cup with small pebbles or a similar substance and the second cup with water. Note that when the two substances are combined, the result is not a full pint. We recommend these activities as a means of stimulating the children to think about parts and wholes and different kinds of mixtures. The lessons can be used as a basis for experiences in forming hypotheses about the ways in which different subtances might be mixed together. These hypotheses can be tested and reformulated if necessary. In each case the instruction should emphasize the part-whole relationship.

8. Help the children to prepare an activity sheet or a bulletin board display on which various parts and wholes are displayed. The task is to draw or to otherwise portray a path connecting a picture of a part to the picture of the appropriate whole.

Fractions

9. Provide the children with two jars of paint—one red, the other yellow—and a number of empty containers. Suggest that in one container the children put some red paint and only a few drops of yellow paint. In another jar, the children should put some yellow paint and only a few drops of red paint. In still another jar, the children should put equal amounts of red paint and yellow paint. Discuss the results. Point out that in every case the same subtances were used, but the results differed because different proportions were used each time. Suggest that the children make their own combinations of red paint and yellow paint. Before making each mixture, the learners should tell how much of each paint will be used and what color and shade is likely to be the result.

10. Display a small container of oil, a teaspoon of sugar, and a container of water. Display also a jar in which there is a mixture of the same ingredients. Ask the learners to identify the parts that can still be seen as separate (water and oil) and the part that cannot be seen any longer (sugar). Focus attention on the fact that certain mixtures can be composed of parts that can still be seen and parts that can no longer be seen.

Interchangeable Parts

Parts of a whole that can be replaced or interchanged with other parts form the basis for other experiences with the part-whole concept. The notion of replaceable or interchangeable parts begins to approximate the traditional concept of fractions. Topics for the development of an understanding of replaceable parts include the tires on an automobile, parts of engines, bicycle tires, light bulbs, batteries, and so on. Essential to these parts is their equivalence in shape and size. Students should be aware of these two conditions. Encouraging them to think of what might happen if something were to be replaced with a part that was the wrong shape and size can aid in the development of this comprehension.

Interchangeable parts of a whole are even more closely related to the concept of a fractional part of a whole. On most cars tires can be rotated to even out the wear on each tire. This is possible because all four tires are usually the same size and shape. Other examples can be cited: some of the drawers in a desk, some buttons on a shirt or a blouse, the pieces of a pie or cake if all the pieces have been cut to the same size.

A discussion of parts that cannot be interchanged can also be beneficial. The socks in a pair of socks can be interchanged; the shoes in a pair of shoes cannot be interchanged. The two lenses in a pair of glasses cannot usually be interchanged. The mittens in a pair of mittens cannot be interchanged. Can the sleeves in a coat be interchanged? Such questions can do much to stimulate children's thinking about like and unlike.

Many mildly handicapped children do not learn to read with any degree of facility until relatively late in their academic careers; therefore, activities such as those just described are valuable in establishing a concept of fractions without requiring the application of skills that have not yet been attained. These same activities are enjoyable ways to help heighten awareness of the many examples of parts and wholes in a child's environment. Although they are not formal mathematics lessons, these activities do provide a readiness for initial work in fractions.

FORMAL INSTRUCTION IN FRACTIONS

Our discussion of formal instruction in fractions will be based on the following premises: Most of us find little use for any but the ordinary fractions in our daily lives, but our benefiting from these requires mastery of both the meaning of fractions and the operations of addition, subtraction, and multiplication. Formal instruction in these areas would begin to take place when a child can be taught to recognize and produce fraction symbols.

The greater popularity of the metric system of measurement has led to some questions about the need for extensive instruction in fractions. There is little evidence to indicate that common fractions will not continue to find applications in many areas of activity—from cooking to carpentry, from sewing to interior decorating. In addition to those applications, the rules for computing with fractions are carried over to algebraic fractions. Many mildly handicapped children pursue college preparatory courses, including algebra, during secondary school. Instruction in fractions, then, should form an integral part of a mathematics program for mildly handicapped children.

Before outlining content and some instructional procedures for fractions, we ask the reader to take a pencil, paper, scissors, some coins, crayons, or whatever. Using these materials, find answers to the following expressions: $\frac{1}{4} + \frac{2}{4}, \frac{1}{3} + \frac{1}{4}, \frac{5}{4} - \frac{3}{4}, \frac{2}{3} - \frac{1}{4}, \frac{1}{2} \times 3, 4, \times \frac{2}{3}, \frac{1}{3} \times \frac{1}{2}, 4 \div \frac{1}{3}, \frac{1}{2} \div \frac{1}{3}$. Did you have trouble finding the answer through manipulatives, even though you could compute the answer? Many of us received instruction in fractions only through symbols; to this day we have difficulty assigning meaning to these symbols and to the algorithms we use. If those of us with average intelligence, better-than-average intelligence, and learning capabilities that are fairly intact cannot assign meaning to content taught only through symbols, we can be sure that such an approach will have little or no lasting value for mildly handicapped children.

The Rational Numbers, Mathematically Speaking

A mathematical definition of a fraction might read as follows: A fraction is a number that can be written as the ratio of two whole numbers. The reader may recall that the whole numbers are 0, 1, 2, 3, and so on to infinity. A mathematical definition of a rational number, then, would be the following: A rational number is one that can be written as the ratio of two integers. The integers include the whole numbers plus a negative partner for every whole number except zero. Clearly, fractions are rational numbers, but the inverse need not be true. That is, $-\frac{2}{3}$ is a rational number but is not a fraction.

These mathematical definitions are nice and clean, but they are somehow foreign to our standard way of thinking about fractions. Menninger (1977, 123) reminded us that "early man, in speech as in writing, drew on his environment for symbols to indicate sizes and quantities; he did not fashion artificial words or symbols to apply only to specific numbers and measures but had no intrinsic connection with them." As evidence of this, Menninger noted that the Babylonians used an image of a measuring vessel, the contents of which were shown to be cut in half, as the symbol for one-half. Fractions originated with the need to provide quantitative labels for parts of wholes. It is that approach with which most of us began our study of fractions.

Fractions can represent parts of a continuous whole or parts of a set composed of discrete elements. In either case, the whole is divided into a number of congruent or equivalent parts. The number of these parts is represented symbolically by the number under the line in a fraction. The number of parts under consideration in any given case is represented by the number above the line.

The congruence or equivalence of the parts into which the whole is broken is a concept essential to the understanding of fractions. Experiences like the following serve to highlight this condition.

UNDERSTANDING FRACTIONS

1. Conduct this activity as a sharing experience. Divide a granola bar or some other edible item equally between two learners. Using a facsimile of a candy bar, repeat the process of cutting in half. Give a child another facsimile and direct him or her to do the same. Discuss the approach the learner chooses to complete the task. Did he or she use a trial and error approach? Did the child measure? Did he or she fold the "bar" in half? What method seems to be the most efficient? the most accurate?

2. Paper-folding activities can be very useful for developing a comprehension of equal parts of a whole and for perceiving the relationships between smaller and larger parts. Make one paper fold to result in two equal parts; make one more fold to divide the paper into four equal parts. Encourage discussion before you make each fold and ask for guesses about how many equal parts will result.

3. With a newspaper extend the concept of dividing a page into more than two equal parts. Show the children how to start by dividing one page into halves; then break each of these halves into halves. Give the children sheets of newspaper and ask them to do the same. After each division, record on the chalkboard the number of parts into which the original sheet of paper has been divided.

4. Help the learners to draw and cut out "happy faces" of different sizes, which may be made from construction paper, newspaper, or any other material at hand. (See Figure 7.1.) Each drawing, when completed, is to be cut in half on the vertical axis or on the horizontal axis. Spread the halves around on a large table. Direct the children to pair halves so that the "happy faces" are complete again. Discuss the relationship between the size of the part and the size of the whole.

Models for Fractions

A perusal of texts and periodicals in mathematics education will result in a collection of several models suitable for teaching fractions. The choice of a model should reflect several concerns. The first of these is the purpose for using manipulative models.

The use of manipulatives in and of itself is not necessarily good or bad.

FIGURE 7.1 Happy faces

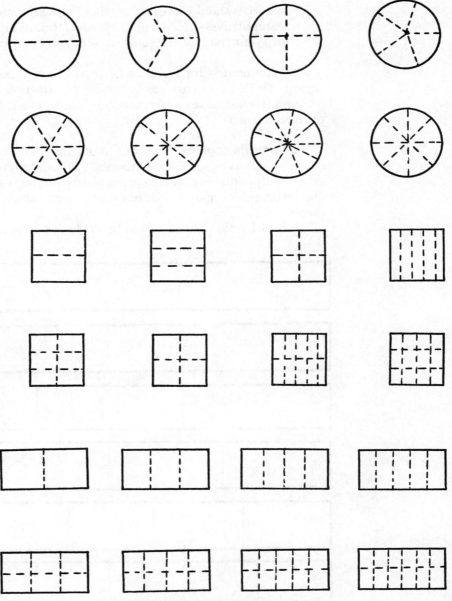

FIGURE 7.2 Geometric regions

Manipulatives serve many purposes: to depict a concept, definition, process, or principle; to justify a process; to provide an algorithmic aid; or to provide instances leading to a generalization. If the materials are used improperly or if they are faulty in some way in and of themselves, they can lead to faulty understanding or an undue reliance on the materials. Some materials are more complex than the content being represented. In every case, an instructor must take care to move beyond the materials to the logical structures being conveyed. Hence we need to take care to observe certain precautions in the choice of manipulative materials:

- They must be appropriate to the concept being developed.
- Materials should be simple with respect to their visual impact, storage, and utilization.
- The materials must be capable of facilitating adherence to mathematical accuracy.

Fractions   **123**

- Materials should be easily obtained or easily made.
- Manipulatives should lend themselves to a multiplicity of uses, both over different levels of the same content and over different content.

Typical models for fractions fall into one of three categories: geometric regions divided into parts, as in Figure 7.2; fraction bars, such as in Figure 7.3; and fraction boxes, which are shown in Figure 7.4. We will examine each of these in turn.

Geometric regions. Most of us are familiar with the pieces of pie frequently used to image fraction names. Certainly circular regions lend themselves easily to division into congruent parts. Equally useful are square regions and rectangular regions. Children should see all kinds as samples of fractional pieces.

Sets of fraction pieces should be made over a period of several weeks by

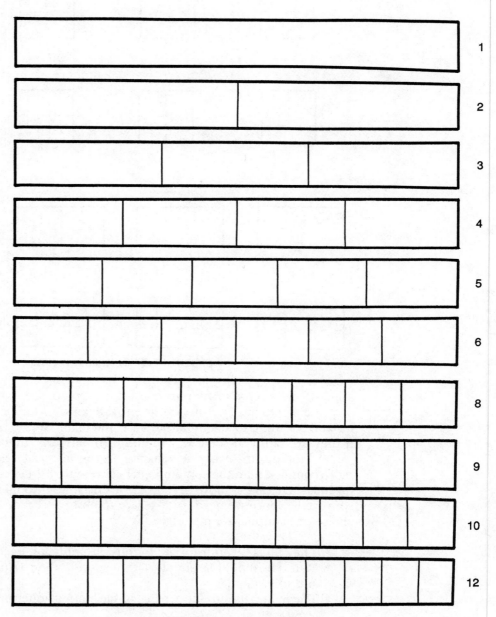

FIGURE 7.3 Fraction bars

the children themselves. During this construction there will be opportunity to discuss each shape, each division, and the relationships between pieces of different sizes; for example, the equivalence between two-fourths and one-half. Teachers have solved the storage problem in various ways, such as using plastic sandwich bags so the contents of each bag are easily seen, or envelopes, or pocket charts. The fraction pieces can be used to provide experiences leading to the invention of the addition and subtraction algorithms for fractions and for justifying a multiplication algorithm and a division algorithm. At each step they provide a means of checking hypotheses about how fractions behave under certain operations.

Fraction bars. As illustrated in Figure 7.3, fraction bars or strips can take several forms. Different bars can be used, each divided to depict different fractions. As a set of materials, fraction strips are easily made and stored.

Relationships between and among different fraction names can be illustrated by placing the strips one under the other. While lending themselves to exploration of addition and subtraction algorithms, fraction strips can also be used for multiplication and division. Also, with little adaptation, fraction bars can be extended to represent a number line that includes both the whole numbers and fractions.

Fraction boxes. Representation by means of fraction boxes (Van de Walle and Thompson 1980) can be efficient. Requiring sheets of paper on which the boxes are drawn and small markers such as dried kidney beans, the model is easily made and stored.

The fraction boxes lend themselves well to the development of addition and subtraction algorithms and, with minor modifications, to an understanding of multiplication and division algorithms. Drawing the boxes on transparent plastic pieces allows for easy discussion of equivalent fractions and their role in computing with fractions. Since the markers indicate only the portion of the whole under consideration, however, the fraction boxes may be considered a more abstract representation.

Before continuing, we encourage the reader to prepare a set of each of the three models previously described. By using these materials as you read the following pages, you will gain experience in the application of the materials, while following the thought processes linking the manipulatives to the concepts being developed.

Basic Fraction Content

A fairly typical scope and sequence of fraction content is outlined in Figure 7.5. Some programs introduce multiplication with fractions before addition and subtraction, since some of the procedures of multiplication find application in the conventional algorithms for addition and subtraction. If children are allowed to "discover" the procedures for addition, those skills of multiplication needed for addition and subtraction will be developed in conjunction with the algorithm. Hence either sequence is appropriate.

At every step of the sequence, mildly handicapped children should be allowed to develop the concepts applicable through repeated experiences with manipulating the materials. As was the case with whole numbers, symbols provide us with a shorthand for writing down what we see with manipulatives and for recording events that would be cumbersome to portray with manipulatives. Only after those two advantages of symbols are fully comprehended should the child advance to using symbols for those events that happen only in our minds.

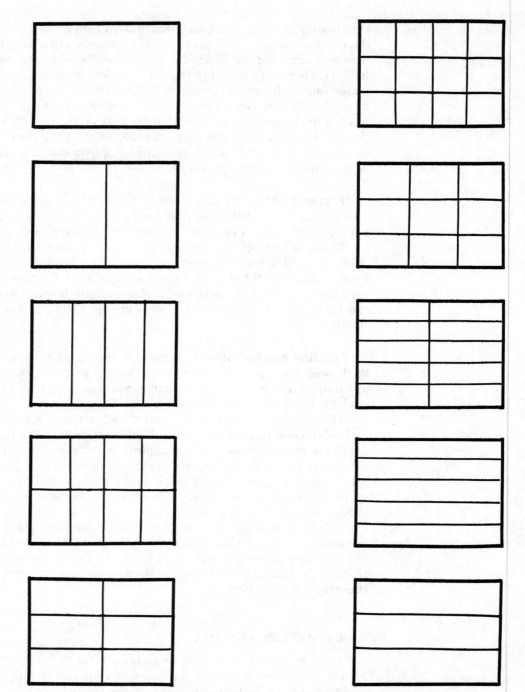

FIGURE 7.4   Fraction boxes

SOME TEACHING SUGGESTIONS

Basic Concepts

As noted previously, each child can participate in the development of a set of fraction pieces for his or her own use. This set should consist of fractional parts of circular regions (at least seven inches in diameter), rectangular regions (8″ × 5″ is a good size), and square rectangular regions (7″ × 7″ is a good size). Fractions with denominators 2, 3, 4, 5, 6, 8, and 10 should find representation. During the making of these sets of pieces much discussion of the fraction names and their meaning can take place. The role of the bottom

Rational numbers named as fractions

1. Basic concepts
   Part/whole
   Numerals
   Role of numerator
   Role of denominator
   Proper fractions
   Improper fractions
   Mixed numerals
2. Addition
   Like denominators, proper fractions
   Like denominators, mixed numerals
   Unlike denominators
3. Subtraction
   Like denominators, proper fractions
   Like denominators, mixed numerals
   Unlike denominators, no renaming necessary
   Unlike denominators, renaming necessary
4. Multiplication
   One factor a whole number
   Both factors proper fractions
5. Division
   Whole number by a proper fraction

Rational numbers named as decimals

1. Basic concepts
2. Addition
3. Subtraction
4. Multiplication
5. Division—whole number divisor

Percent

1. Basic concepts
2. Finding a percent of a number

FIGURE 7.5 Scope and sequence for rational numbers

number (denominator) and the top number (numerator) should be understood. Also necessary is the awareness that the denominator is nothing more than a label.

Each fraction piece should be labeled on one side only with its fraction name. Then students can employ the pieces for both nonsymbolic activities and symbolic activities when appropriate. The time spent in making these fraction sets will pay dividends later.

Mildly handicapped children can also participate in the construction of fraction bars. Each child should have a set of ten bars drawn on paper, each divided as illustrated in Figure 7.3. Accompanying the paper strips should be one or more identical sets of strips drawn on transparent plastic. Each part of each strip should be labeled with its fraction name. As with the fraction pieces, children can work with the strips to develop an appreciation of the meaning of the fraction names and the relationships among them.

Fraction boxes should be prepared in a manner similar to that of the fraction bars. The rules for using the markers (e.g., dried kidney beans or small plastic chips) must be explained to the youngsters. A marker placed in a portion of a fraction box indicates the shading of that portion. Once that representation is understood, the fraction boxes can play a role in the development of the basic concepts.

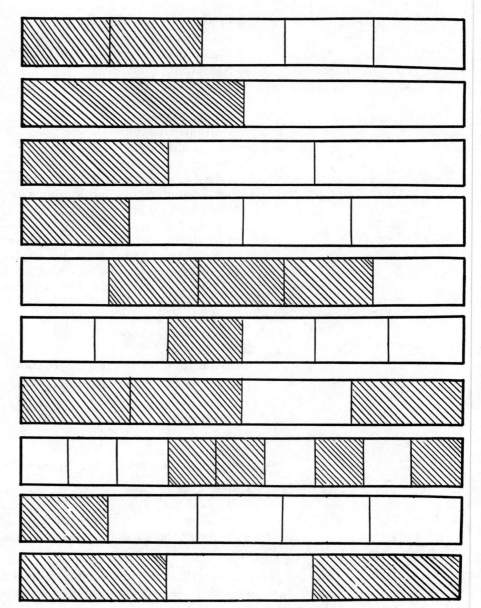

FIGURE 7.6 Display sheet

The activities that follow are designed to provide the reader with some strategies for ensuring that mildly handicapped children understand the meaning of proper fractions, representation of the whole, and what we traditionally call improper fractions and mixed numerals. Each activity is presented in terms of one of the materials previously discussed. We encourage the reader to modify the activity by using other materials.

During the activities, stress (frequently) the role of the denominator as a label, for labeling is a function the denominator performs. Familiarity with this interpretation of the denominator will facilitate the acquisition of rules for adding, subtracting, and dividing fractions.

A C T I V I T Y

UNDERSTANDING PROPER FRACTIONS, IMPROPER FRACTIONS, MIXED NUMERALS, AND REPRESENTATION OF THE WHOLE

1. Using a set of circular-region fraction pieces, lay out a representation of ¾. Ask the learners to do what you did. Repeat for representations of several

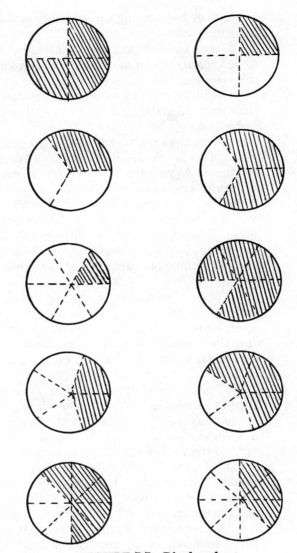

FIGURE 7.7 Display sheet

proper fractions. In each case the children are to imitate your behavior. Finally, put out a representation of $\frac{5}{6}$ using circular-region fraction pieces. Ask the children to switch to parts of rectangular regions to illustrate the same fraction.

2. For this activity sheet you will need colored pieces of transparent plastic, which are cut to the same width as the fraction bars, a display or activity sheet like that illustrated in Figure 7.6, and a set of fraction strips. If using the display, place it so that each child can see the figures clearly. Otherwise give each child a copy of the activity sheet. Use the colored transparent pieces to shade in a portion of a fraction strip to depict a proper fraction; for example, $\frac{2}{3}$, by covering two of the three divisions on the strip divided into thirds. Ask the children to find the representation of what you have done on the display or activity sheets. Repeat for the different proper fractions represented on the display. After the children have gained facility in pointing to the representation of what you are showing, distribute a copy of an activity sheet or prepare a display like that shown in Figure 7.7. Repeat the activity, but this time the children are required to recognize the same proportions in shapes that are different from the strips. Note: Many mildly handicapped children may find selecting the correct item from ten choices too difficult. In such a case, the instructor can limit the number of alternatives by using smaller

Fractions

displays, or by covering up some of the display for a particular student or a particular task.

3. With fraction boxes and small chips, portray different proper fractions. As you place each set of chips, ask the children to name aloud the fraction you have represented.

4. With fraction pieces, represent different proper fractions. As you complete each depiction, ask a child to label the representation with its written name.

5. Display representations of proper fractions, using Figure 7.6 as a model. Provide each child with sets of fraction pieces, fraction strips, or fraction boxes. As you point to each representation in the display, ask the children to use the manipulatives to make a representation of the same fraction.

6. Present displays like those shown in Figures 7.6 and 7.7 again. As you point to a representation in one display, the children are to find the representation of the same fraction in the other display. Note: Once again, precautions may be necessary for some students. This task may be modified by showing a child only one model at a time and restricting the number of choices to two or three.

7. Show the children a representation of a fraction. Ask the learners to name the fraction being represented.

8. Show the children a representation of a fraction. Ask the learners to write the fraction being represented. Note: This activity can take the form of a worksheet, examples of which can be found in many textbooks and workbooks.

9. Provide the children with fraction boxes and small chips. As you name a fraction aloud, the children are to depict that fraction.

10. Display representations of several proper fractions, again using Figures 7.6 and 7.7 as guides. As you name a fraction aloud, the children are to point to the representation of that fraction.

11. Repeat Activity 9, but instead of naming fractions aloud, show the children the fraction written on the chalkboard.

12. Repeat Activity 10, but instead of naming fractions aloud, show the children the fraction written on the chalkboard. Note: Activities 11 and 12 can easily be accomplished through worksheets. Before assigning such worksheets, however, group activities or activities conducted on a one-to-one basis can help to ensure that children understand the task and that they will not be practicing inappropriate behaviors.

We have taken the time and space to spell out the twelve activities because they follow a pattern that can be used over and over again as new content in fractions is pursued. Hence, in the twelve activities just described, the substitution of *improper* for *proper* will provide another set of tasks to ensure a grasp of improper fractions. With minor adaptations, the activities can be used to teach the mixed-numeral notation for improper fractions named as wholes plus parts of wholes. By mastering the combinations described in the activities, the instructor is in effect putting himself or herself in possession of a set of strategies that can be transferred to many different concepts.

At some point in the building of these basic concepts, the number *one* as a denominator can be introduced. When the whole is not divided into any parts, there is only *one* part. Hence 5 can be written as $\frac{5}{1}$. This simple insight is helpful not only for arithmetic with fractions but also for later work in algebra.

Using fractions to name parts of sets of discrete elements can be a more difficult concept than naming congruent parts of a continuous whole. Finding one-fourth of a set of four is relatively easy; dividing a set of twelve into fourths is more difficult. Students can apply their knowledge of division to

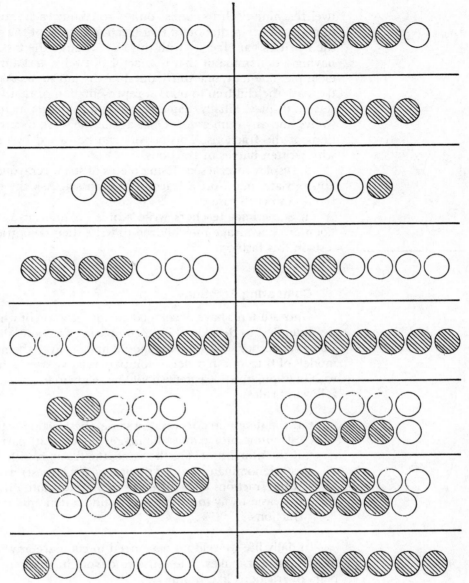

FIGURE 7.8   Activity sheet

exercises that involve finding equivalent parts of sets. Activities like the following may help to develop this more sophisticated discernment. Alternatively, the instructor may wish to postpone finding parts of sets until multiplication is taught.

NAMING PARTS OF SETS

1. Explain to the learners that subsets can be considered as parts of the entire set. To name the fraction we must count all the members of the set to find the denominator and then count the members of the subset to name the numerator. Distribute copies of an activity sheet like that illustrated in Figure 7.8. For each frame, discuss the size of the shaded subset and the size of the entire set. Ask the learners to write the fraction represented in each case.

2. Display several representations of a set of twelve items. Begin by naming fractions with the denominator twelve; for example, three-twelfths, seven-

Fractions   **131**

twelfths, nine-twelfths, and so on. As you name each fraction, the children are to mark a representation of that fraction on one of the sets on display. When the learners gain facility with this task, change the task by naming fractions having a denominator that is a factor of twelve and a numerator of one; for example, one-half, one-third, one-fourth, and so on. As you name each fraction, ask the children to mark a representation of that fraction on one of the sets on display. Finally, name fractions with denominators that are factors of twelve and any numerator. Once again, the learners are to mark representations of the fractions. This activity can be varied by providing the learners with written names of fractions.

3. Display several sets. Using pieces of string, crayon, chalk, or whatever is appropriate, mark off a fraction of each set. Ask the children to name the fraction in each case.

4. Sometimes teachers write scores on tests or assignments in fraction notation. Encourage the children to write their own grades on homework or tests in this fashion.

Comparing Fractions

Once students have developed an appreciation of what a fraction is, activities intended to develop skill in comparing fractions should be undertaken. Like every other skill, students should initially avail themselves of one or more models of fractions in order to compare relative sizes. After many experiences with manipulatives, learners should be ready to formulate, with guidance, the following rules:

- If the denominators are the same, the fraction with the larger (or largest) numerator names the bigger (or biggest) part.
- If the numerators are the same, the fraction with the smaller (or smallest) denominator names the larger (or largest) part.
- If the fractions have different numerators and different denominators, the easiest way to compare is to find a denominator common to all the fractions.

In daily life we seldom have need to compare any but the simplest fractions. Hence it makes little sense to devote instructional time to comparing pairs of fractions like $\frac{7}{11}$ and $\frac{13}{19}$.

Equivalence Classes

An important concept both for a mastery of fractions and a mastery of the operations we perform on fractions is that of equivalent fractions. Each and every fraction can be named by an infinite number of fraction symbols. Each of these groups of names is called an *equivalence class*. We usually name an equivalence class by the simplest name for the fractions it represents. Thus, the equivalence class for $\frac{1}{2}$ includes $\frac{2}{4}$, $\frac{3}{6}$, $\frac{4}{8}$, and so on. Each new member of an equivalence is obtained by multiplying or dividing both the numerator and denominator of a fraction by the same number. That is, $\frac{12}{24} = \frac{1}{2} \times \frac{12}{12}$.

Equivalence classes are easily introduced during the construction of models for fractions. For example, after the children have completed the models for halves and fourths, the relationship between fourths and halves can be demonstrated, and the sentence summarizing that relationship can be recorded: $\frac{1}{2} = \frac{2}{4}$. When using fraction pieces the equivalences can be demonstrated by superimposing sets of pieces on other pieces; for example, placing three one-sixth pieces on top of a one-half piece or on top of two one-fourth pieces. When using fraction strips or the boxes, the equivalences can be shown by placing transparent strips or boxes on top of other strips or boxes. By scrupulously recording the equivalences, each student will have lists of equivalent fractions for future reference. Some mildly handicapped children may go beyond these lists to formulate the rule for obtaining equivalent fractions. Ideally, this formulation should come about as a result of experimenting with the models.

Some mildly handicapped children may need to go a step further. Instead of being able only to find a fraction equivalent to another one, we sometimes need to make a judgment about the equivalence of a pair of fractions. By induction, youngsters can be taught the rule for making such decisions: If $a/b = c/d$, then the product of a and d is equal to the product of b and c. The rule works the other way also; if $ad = bc$, then $a/b = c/d$. Mildly handicapped children who progress to algebra courses will see that rule again.

In summary, then, the concepts and skills youngsters should have before their introduction to the operations with fractions include: (1) an understanding that fractions name and can be represented by parts of continuous quantities or sets of discrete elements; (2) skill in writing fraction names and in representing fractions by way of manipulatives and pictures; (3) facility in comparing fractions; (4) the ability to write improper fractions as mixed numerals and vice versa; and (5) skill finding other members of the equivalence class to which a fraction belongs.

Addition

When the child's world of numbers is extended to include fractions, this extension carries with it the need to perform the operations of addition, subtraction, multiplication, and division in a manner that is sometimes different from and sometimes similar to these same operations on whole numbers. Care must be taken from the beginning to ensure that faulty behaviors do not develop. By allowing the children to experiment with the models and to invent their own procedures, we can do much to encourage the comprehension that the algorithms are a shorthand way of recording what is happening.

Representations of addition with fractions can be achieved using any of the models previously described. When the fractions to be added have the same denominator, the representation is simple. A representation of one addend can be placed next to the representation of the other. By placing a fraction strip above the representation of the two addends, the sum can be found, as illustrated in Figure 7.9.

The learners should find the answers to several examples and record both

Example: $\dfrac{1}{5} + \dfrac{3}{5}$

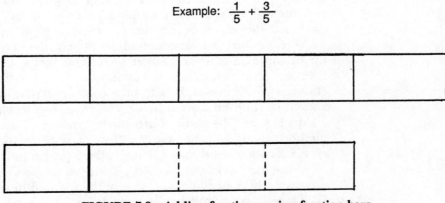

FIGURE 7.9 Adding fractions, using fraction bars

the expression and the sum. After several examples have been completed, the children should be encouraged to develop a rule. Previously we suggested that the students become familiar with interpreting the denominator as a label. Since it is such, it doesn't change. Just as three pounds and four pounds would become seven pounds, three-eighths plus four-eighths would equal seven-eighths. If the sum turns out to name an improper fraction, we usually rename it as a mixed numeral.

By carefully selecting examples, we can introduce adding fractions with unlike denominators quite simply. The procedure will be outlined with fraction strips. Either of the other two models can be used just as effectively.

Suggest that the children represent the addition of $\frac{1}{2}$ and $\frac{1}{4}$. By now they should be adept at the representation. However, what they show is still $\frac{1}{2}$ and $\frac{1}{4}$. Explain that it is more *efficient* to name the sum with one fraction. Ask the children to experiment with other fraction strips to see if the length matches a length on any one strip. Through trial and error, the children should discover a match using the strip divided into fourths, the strip divided into eighths, and the strip divided into twelfths. Write the sentences evident from each match: $\frac{1}{2} + \frac{1}{4} = \frac{3}{4}$; $\frac{2}{4} + \frac{1}{4} = \frac{3}{4}$; $\frac{4}{8} + \frac{2}{8} = \frac{6}{8}$; $\frac{6}{12} + \frac{3}{12} = \frac{9}{12}$. The answers name equivalent fractions. The simplest of these is $\frac{3}{4}$.

Repeat the exercise many times, using combinations such as $\frac{1}{2} + \frac{1}{3}$, $\frac{3}{4} + \frac{2}{3}$, and so on. Record the sentences for each solution, record the denominators in each addend and the denominator in the sum each time. After several examples have been finished, ask the children if they see any connection between the denominators of the addends and the denominator of the sum. Elicit the generalization that in each case the denominators of the addends are factors of the denominator of the sum. Hence, each addend can be renamed in terms of a common denominator.

For many mildly handicapped children, it is a difficult task to find a common denominator when the fractions to be added have unlike denominators. Skill in finding a common denominator finds application not only in the addition of fractions, but also in subtraction and division. For that reason, time spent in developing mastery of that skill is time well spent. We shall discuss some computational methods for finding common denominators, although some mildly handicapped children may never progress beyond examining sets of equivalent fractions to find a common denominator.

The easiest strategy is that of finding the product of the numbers under consideration. A common multiple of 3 and 4 is 12, the product of 3 and 4. A common multiple of 3 and 6 is 18; a common multiple of 6 and 8 is 48. These common multiples are not necessarily the least of the common multiples for some of the pairs involved, but common multiples they are. They serve the purpose of common denominators quite well. If necessary, the learner can rename the sum in its simplest form.

Another method requires the listing of the multiplies of each of the numbers for which a common multiple is necessary. From inspection, the student can select the lowest or the least common multiple.

For finding the least common multiple or the lowest common denominator of common fractions, either of the two methods outlined above is relatively efficient and effective. For most of us, including mildly handicapped children, these methods will suffice for everyday use. Hence instruction should be directed toward mastery of those two techniques.

In Figure 7.10 we have provided examples of the two processes and the method relying on prime factors. More capable mildly handicapped children, especially those who may pursue college preparatory programs, will profit from instruction in the method relying on prime factors. Most texts in elementary school mathematics present the method in fuller detail.

Once a common denominator has been obtained, the child is faced with

	Method 1: Product of denominators	Method 2: Finding multiples	Method 3: Prime factors
Example $\frac{1}{4} + \frac{1}{3}$	$4 \times 3 = \text{⑫}$	4,8,⑫, 16 3,6,9,⑫	$2 \times 2 \times 3 = \text{⑫}$

FIGURE 7.10 Finding common denominators

the task of renaming the addends in terms of that common denominator. This skill involves no more than finding fractions equivalent to the addends in the original example. In each case, once the common denominator is known, the learner must divide the new denominator by the former one, to find the factor by which the former denominator had been multiplied. The numerator of the equivalent fraction must be the product of the former numerator and the factor resulting from the division. Encouraging the children to illustrate each step with manipulatives will help to prohibit the rote memorization of a series of meaningless steps.

Some cautions are in order here. Fractions are written in the form of a ratio or a comparison of two numbers—one representing the number of parts into which a quantity has been divided, and the other naming the number of parts under consideration. Generally speaking, we humans prefer to deal with smaller numbers than with larger numbers. For this reason we usually write fractions in their simplest form. Two out of three is easier for us to attach a meaning to than 32 out of 48 or 90 out of 135. This cognitive preference has translated itself into the practice of naming fractions in their simplest form whenever feasible, the process sometimes referred to as *reducing fractions*. To the same end we usually rewrite improper fractions as a mixture of whole numbers and proper fractions. Instruction in reducing fractions or in writing improper fractions as mixed numerals should reflect this purpose. We suggest that children who forget or who do not choose to rewrite their answers to computation exercises should not be penalized for this neglect to the extent that their successful completion of the example is unrewarded.

When discussing the addition of whole numbers, we noted that the operations of arithmetic are performed on numbers, not things. We repeat that principle here and offer some evidence that further illustrates the point. In mathematics, the expression $\frac{1}{2} + \frac{1}{2} = 1$ is true. Imagine someone making hot cereal, following a recipe on the box, which calls for one-half cup of water to be mixed with one-half cup of the cereal. Imagine further a desire to be efficient in the task of filling a measuring cup to the half-cup level with water and then filling the cup to the full-cup line with hot cereal. Such a mixture would result in a very dry dish of hot cereal. We suggest that the instructor discuss this and similar situations with the children.

Subtraction

Much of what was written about adding fractions applies to the subtraction of fractions. We urge the reader to turn to the preceding chapter of this book and review the section that dealt with the three types of applications that call for subtraction. We recommend that these three types of situations be reviewed with the children and that they be illustrated with representations of fractions.

In many cases, instruction in adding fractions can be accompanied by the teaching of subtraction, since the skills with respect to working with or without common denominators are similar. Subtraction, however, does require one skill not found in the addition sequence—namely, renaming the top fraction

Fractions **135**

$$7\tfrac{1}{2} \;=\; 7\tfrac{2}{4} \;=\; 6\tfrac{6}{4} \;=\; 6\tfrac{6}{4}$$
$$4\tfrac{3}{4} \;=\; 4\tfrac{3}{4} \;=\; 4\tfrac{3}{4} \;=\; 4\tfrac{3}{4}$$
$$2\tfrac{3}{4}$$

Step 1   Step 2   Step 3   Step 4

FIGURE 7.11 Subtraction example

in an example when the bottom fraction has a numerator greater than that of the top fraction, after both fractions have been renamed in terms of a common denominator. Two methods for doing this can be mentioned, although once again we would recommend presenting such an example to a learner as a problem-solving task to be accomplished with manipulatives. By watching the direction taken by the child, the instructor may see which method comes more naturally to the learner.

If the groundwork has been laid with respect to renaming in base ten, renaming with fractions becomes an extension of the same skill. The difference lies in renaming a *one* according to the denominator of the fraction under consideration. In the exercise shown in Figure 7.11, the transition from Step 2 to Step 3 consists of no more than renaming one of the 7 *ones* as $\tfrac{4}{4}$, and then adding the $\tfrac{4}{4}$ to the $\tfrac{2}{4}$. Note: At each step after the first, the entire mixed numeral is written. The sentence $7\tfrac{1}{2} = \tfrac{2}{4}$ is a false one. Students should always be required to write only true number sentences.

Some students or instructors prefer an alternate approach. In this strategy, both the top fraction and the bottom fraction are written as improper fractions either before or after a common denominator has been found. As a result no renaming is necessary.

Multiplication

Many mathematics programs introduce multiplication with fractions before treating addition and subtraction with fractions. The rationale for that procedure lies in the simplicity of the algorithm for multiplication, once students know how to rename mixed numerals as improper fractions. Our choice of sequence in this chapter reflects no more than a desire to be consistent with the sequence chosen for whole numbers. In actual practice, either sequence has desirable features.

The traditional multiplication algorithm for fractions reflects some properties of real numbers rather than a reasonable shorthand for writing down what happens—that is, what we see—when we manipulate fraction models. Nonetheless, employing materials is necessary for students to acquire both a grasp of what a multiplication expression means and a rationale for the algorithm. The following activities will provide some ideas for representing multiplication expressions.

A C T I V I T Y

UNDERSTANDING THE MULTIPLICATION OF FRACTIONS

1. Provide each child with a carton designed to hold one dozen eggs; also provide each child with some small chips. Count the egg holders in the carton. Ask the children to put chips into three-twelfths of the egg holders. Provide guidance if necessary. Repeat the request for different proper fractions with the denominator twelve. Then challenge the learners to solve these problems: put chips into $\tfrac{1}{2}$ of the egg holders, $\tfrac{1}{3}$ of the egg holders, $\tfrac{3}{4}$ of the egg holders. Observe the efforts of the children to solve the problems. After each problem, discuss the approaches taken and write the number sentence that has been represented: $\tfrac{1}{2} \times 12 = 6$, $\tfrac{1}{3} \times 12 = 4$, $\tfrac{3}{4} \times 12 = 9$.

×	1	2	3	4	5	6	7	8	9	10	11	12	13	14	15	16	17	18	19	20
$\frac{1}{2}$																				
$\frac{1}{3}$																				
$\frac{1}{4}$																				
$\frac{1}{5}$																				
$\frac{1}{6}$																				

FIGURE 7.12 Multiplication chart

2. Provide a geoboard and rubber bands of different colors. Show the children how to depict an expression such as $\frac{1}{3} \times 6$ by enclosing 6 pegs or nails with a rubber band of one color and then enclosing $\frac{1}{3}$ of those pegs or nails with a rubber band of another color. Repeat for several expressions.

3. Ask the youngsters to collect and to bring to class pictures of items agreed upon by the group: popular music stars, types of food, sports personalities, or the like. Arrange sets of these pictures on a suitable surface. With yarn, strips of paper, crayon, or whatever, set off a fraction of each set. To the right of the sets, arrange in random order the multiplication expressions represented by the parts of the sets in the display. Each learner, in turn, is to make a path between each set and the expression represented by that set. The activity can also be accomplished through a worksheet. One youngster can make up a worksheet for another child by drawing representations of multiplication expressions in one column and writing the expressions represented in random order in a second column.

4. Provide each learner with a copy of a recipe. The learners' task is to double the recipe or to triple it. The youngsters can write the quantities needed in each case, draw pictures representing the quantities needed, or both. At another time the learners' task can become that of finding one-half of the amount of each ingredient.

5. Provide each learner with a copy of the chart presented in Figure 7.12. The children are to complete the chart by writing only those answers that are whole numbers. When the chart is completed, discuss the results. Note that products are whole numbers only when the whole-number factor is a multiple of the denominator of the fraction that is the other factor of the multiplication expression; for example, $\frac{1}{2}$ of any even number is a whole number, $\frac{2}{3}$ of any multiple of three is a whole number, and so on.

Discussion about the results of the activities just described should help the learners to observe that multiplying a whole number by a fraction is equivalent to dividing the whole number by the denominator of the fraction and multiplying the result by the numerator. That is, $\frac{2}{3} \times 12 = (12/3) \times 2$. When we do a fraction example, we usually write this as $(12 \times 2)/3$. When the numerator is one, multiplying a whole number by a fraction gives the same result as dividing the whole number by the denominator of the fraction. Although this is apparently a simple concept, many youngsters never make the connection for themselves, yet those mildly handicapped children going on to algebra will need that understanding.

In Figure 7.13 we illustrate models for representing other types of multiplication expressions. Instructional time should be devoted to helping children represent the different kinds of expressions, each time noting the role of the numerators and denominators in the construction and in the result. The last expression in the series most closely resembles the cross-product interpretation of multiplication. The numerator of the product is derived by "crossing" the numerator of one factor with the numerator of the other factor. The

Fractions

Expression	Interpretation	Representation
$6 \times \frac{1}{2}$	6 pieces of $\frac{1}{2}$	
$\frac{1}{2} \times 6$	$\frac{1}{2}$ of a set of 6	
$\frac{1}{2} \times \frac{1}{3}$	$\frac{1}{2}$ of $\frac{1}{3}$	
$\frac{1}{3} \times \frac{1}{2}$	$\frac{1}{3}$ of $\frac{1}{2}$	

FIGURE 7.13 Models for multiplication expressions

denominators play the same roles. Thus the algorithm for finding the product of two fractions $a/b \times c/d$ is expressed as $(a \times c)/(b \times d)$. Having derived this algorithm, the children can observe that it works for all kinds of multiplication expressions. When one or more factors is a mixed numeral, the computation becomes easier if we first change the notation into that of an improper fraction. By dividing factors common to numerators and denominators, we can sometimes make the multiplication easier.

At this point it may be valuable to discuss with the children the similarities between the rules for fractions and those for whole numbers. $5 + 3$ could be written as $\frac{5}{1} + \frac{3}{1}$; 5×3 could be written as $\frac{5}{1} \times \frac{3}{1}$. Because the number *one* acts as an identity element for division and multiplication (it leaves the other factor unchanged), we usually do not write it when we work with whole numbers. A discussion of this nature can help children to realize the connections among the different topics of arithmetic and the basic unit of the rational-number system.

Division

Division with fractions finds relatively little application in daily life. The most commonly seen type of example is the division of a whole number by a fraction. For example, ten yards of fabric have been donated to the local nursery school. If one puppet requires one-half yard of material, how many puppets can be made from the donated cloth? An example of measurement division, the problem is solved easily by cutting the cloth into half-yard pieces and counting the pieces.

The division sentence $10 \div \frac{1}{2} = ?$ corresponds to the multiplication expression $? \times \frac{1}{2} = 10$, or $\frac{1}{2} \times ? = 10$. The first expression asks how many halves there are in 10; the second asks one-half of what number is 10. Likewise, the expression $\frac{1}{2} - \frac{1}{3} = ?$ can be interpreted as asking how many pieces that are $\frac{1}{3}$ in size there are in $\frac{1}{2}$; or, $\frac{1}{3}$ of what number is $\frac{1}{2}$? If a division algorithm is to be taught, mildly handicapped children should have many experiences in interpreting the expressions and finding answers through manipulatives. Only after such experiments will the results seem reasonable. For example, when we ask how many pieces that are $\frac{1}{3}$ in size will be found in $\frac{1}{2}$, we begin to realize that the answer must be a little more than one. We can estimate how *much* more through the manipulatives.

The conventional algorithm for division with fractions requires the reciprocal of the divisor and the application of the previously learned algorithm for multiplication with fractions. That the rule seldom makes any sense to its learners is concluded from the rapidity with which most of them forget it or, at a minimum, forget which of the two fractions is to be inverted. We suggest

Example: $\frac{1}{3} \div \frac{1}{4}$

A common denominator is 12.

$\frac{1}{3} \div \frac{1}{4} = \frac{4}{12} \div \frac{3}{12}$, which is the same as $4 \div 3$, or $\frac{4}{3}$ or $1\frac{1}{3}$.

FIGURE 7.14   Division algorithm—common denominator

an alternate algorithm, one that relies on the learner's previous experience with division of whole numbers. Examine and, where necessary, complete the following number sentences:

$$6 \div 2 = 3$$
$$60 \div 20 = 3$$
$$600 \div 200 = 3$$
$$.6 \div .2 = 3$$
$$.06 \div .02 = \underline{\hspace{2cm}}$$
$$60{,}000 \div 20{,}000 = \underline{\hspace{2cm}}$$
$$600{,}000 \div 200{,}000 = \underline{\hspace{2cm}}$$
$$6\#\# \div 2\#\# = \underline{\hspace{2cm}}$$
$$6/14 \div 2/14 = \underline{\hspace{2cm}}$$

The rule illustrated above, inductively, can be simply stated by saying that if 6 of any number is divided by 2 of any number the result will always be three. More formally stated, $6n \div 2n = 3$ for any value of n except zero.

For mildly handicapped children who have had good instruction in division, translating an example of division with fractions into one with whole numbers is a beneficial strategy. The procedure is outlined in Figure 7.14.

Decimals

Our base-ten system of notation produces an alternate set of symbols to designate rational numbers having denominators that are powers of ten; for example, $\frac{3}{10}$, $\frac{4}{1000}$. In the numeral 1111, each digit represents one-tenth of the value of the digit to its left. A logical step, then, when locating the digit representing the unit, is to place some symbol to note that digit and to write a digit representing one-tenth of the unit to the right of that mark. In the United States, it is customary to mark a point to highlight the units place. The *decimal point*, then, simply tells us where the unit is located. The unit can change. For example, in the phrase 3.9 million, the unit is recognized to be one million.

The simplicity of the concept of using place value to designate fractional parts can be offset by the difficulty of applying the concept to reading decimal fractions. The best way to avoid difficulty is to provide youngsters with representations of decimal fractions representing denominators of ten, one hundred, and one thousand. Many instructors provide students with mnemonic devices to help them translate the symbols when necessary. Such aids can be helpful if they are not used as a substitute for instruction in the meaning of the notation.

Computing with mixed numerals written in decimal notation relies on the same processes required for computing with whole numbers, since those procedures are based on our decimal number system. Children should be allowed to experiment with representations of decimals to convince themselves that the rules work in this context. They will readily observe that multiplication requires one more precaution. They should be able to demonstrate that $\frac{1}{10} \times \frac{1}{10} = \frac{1}{100}$, $\frac{1}{10} \times \frac{1}{100} = \frac{1}{1000}$, and so on. Hence, $.1 \times .1 = .01$; $.1 \times .01 = .001$.

Through those examples the simple rule can be developed: The number of *decimal places* in the answer must be the sum of the number of decimal places in the factors. Any zeros in the rightmost columns can be eliminated.

In daily life we seldom need decimal fractions beyond those naming thousandths. In those areas requiring decimal fractions that name very small parts, scientific notation is frequently employed. Mildly handicapped students pursuing such topics will require instruction. Suggestion and appropriate sequences can be found in most junior high school mathematics textbooks.

All rational numbers have a decimal-fraction name. The decimal name can be derived by dividing the numerator by the denominator. Sometimes the complete name can only be indicated. For example, $\frac{1}{3}$ in decimal notation would be $0.333333333\ldots$; $\frac{1}{7}$ would be $0.142857142857142857\ldots$ To avoid unnecessary writing, we conventionally place a bar over the sequence of digits that repeats itself. That is, $\frac{1}{3} = 0.\overline{3}$; $\frac{1}{7} = 0.\overline{142857}$. In applied contexts, we usually round off to two or three places.

All decimal fractions, however, do not have a rational-number representation. Mildly handicapped children moving on to algebra courses will meet the irrational numbers. While instruction in the nature of the irrational numbers may not be appropriate for many mildly handicapped children, they should be aware of the incompleteness of the rational-number system. This can be indicated quite simply by a reference to other kinds of numbers that we don't need to study now.

Rounding Off

Mildly handicapped students pursuing vocational educational programs will need to master the skill of rounding off numbers. Some discussion will convince students that exact measures are seldom necessary. Even Uncle Sam allows us to round off tax-form entries to the nearest dollar. Students should be encouraged to express suggestions for degrees of accuracy necessary in different situations.

Representation of parts of number lines can be helpful in introducing the rounding off procedure. We "round off" by moving to the multiple of ten, hundred, thousand, tenth, hundredth, and thousandth that is nearest to where the number falls on the number line. The multiple of ten chosen is dictated by the degree of accuracy desired. By agreement, we consider 5 to be closer to the next highest multiple.

The rule follows readily: Mark the digit in the place representing the desired degree of accuracy. Look at the value of the digit to the right of that place. If that digit is 5 or greater, increase the marked digit by one. Otherwise, do not change the value of the marked digit.

Because the content under examination here is sometimes covered relatively late in the academic experience of a mildly handicapped child, the temptation to rely on symbolic representation is strong. Such reliance may serve to destroy what has been carefully constructed previously. Using manipulatives to build comprehension will effect more efficient learning in the end.

Percent

A difficult concept, *percent* is a construct that finds many applications in our lives. It is, in a sense, the basic statistic, serving to summarize the relationships between and among parts of many types of quantity. The use of percent divides the whole, no matter what it is, into one hundred parts, and allows us to examine some number of those parts. The quantities divided may be groups of people, prices of merchandise, essential nutrients in our food, components of a metallic substance, and so on. A basic grasp of percent is even necessary to read the daily newspaper intelligently.

We strongly recommend that teachers spend time developing the concept of percent with mildly handicapped children. The notion of *one hundred* as the denominator for every number followed by a percent sign is important. To build this knowledge, ten-by-ten grids will be useful. The following set of activities provides suggestions for using such grids. The grids can be easily made on graph paper. The size of the unit blocks can vary from one-quarter-inch blocks to one-centimeter blocks.

A C T I V I T Y

UNDERSTANDING THE CONCEPT OF PERCENT

1. Provide the learner and yourself with a ten-by-ten grid and small chips. Fill in a portion of the grid with the chips. Ask the learner to do the same. Repeat for different portions of the grid.

2. Display several pictures showing filled portions of a ten-by-ten grid. Use small chips to fill in a portion of a ten-by-ten grid. Ask the learner to point to the picture that shows the same portion shaded in.

3. Fill in a portion of a ten-by-ten grid with small chips. Ask the learner to name the number of squares you have filled in and to name the total number of squares. Repeat the activity, asking the learner to name the filled-in portion using a fraction with denominator 100.

4. Fill in a portion of a ten-by-ten grid. Ask the learner to write the fraction naming the portion of the grid that has been filled in.

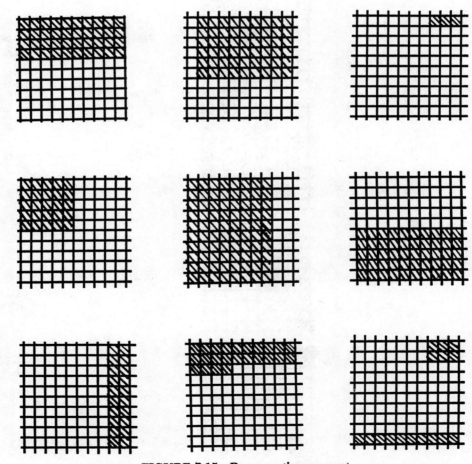

FIGURE 7.15   Representing percent

Fractions   **141**

5. Provide the learner with an activity sheet like that shown in Figure 7.15. Use small chips to fill in a portion of a grid equal to the shaded portion of a grid on the activity sheet. Ask the learner to find the grid showing the same portion filled in.

6. Display a picture of a grid with a shaded portion equal to the shaded portion of one of the grids on the activity sheet (Figure 7.15). Ask the student to find the matching grid on the activity sheet. Note: the portion shaded in the display need not have the same configuration as that shown on the activity sheet. Also, the grid need not be ten-by-ten once the student gains some familiarity with the task.

7. Ask the student to name aloud the fraction represented by each grid on the activity sheet, always using 100 as the denominator.

8. Ask the student to write the fraction represented by each grid on the activity sheet, again always using 100 as the denominator.

9. Provide the student with a ten-by-ten grid and some chips. Explain that

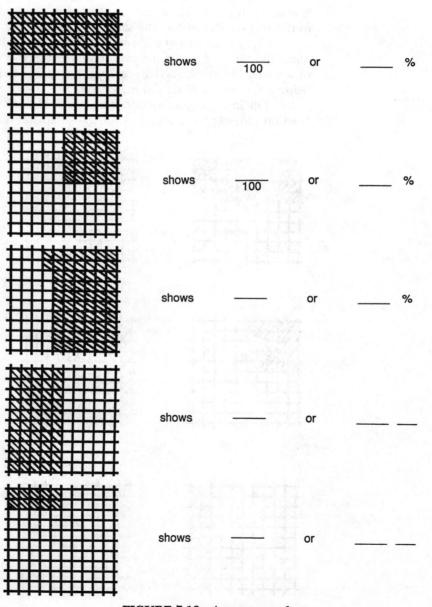

FIGURE 7.16   A percent task

Chapter Seven

the word *percent* means we are using a fraction with a denominator of 100. The numerator is the number we say before we say percent. Dictate several expressions using the word *percent*. Ask the learner to represent the fractions on the grid.

10. Direct the student's attention once again to the activity sheet (Figure 7.15). Name a percent represented by one of the grids. Ask the learner to point to the appropriate grid. Repeat for other grids.

11. Dictate several percent expressions. Ask the child to describe how to write each expression as a fraction and as a decimal fraction.

12. Dictate several percent expressions. Ask the learner to write each expression as a fraction and as a decimal fraction.

13. Introduce the symbol for percent. Dictate several fractions with denominators equal to 100. Ask the learner to write the percent expression.

14. Write a percent expression on the chalkboard. Ask the student to picture the expression on a grid.

15. Provide the learner with a copy of an activity sheet like that shown in Figure 7.16. Ask the student to complete the sheet.

16. Write a percent expression on the chalkboard. Ask the learner to describe how to write the expression as a fraction and as a decimal.

17. Provide the learner with activity sheets like those shown in Figure 7.17. Ask the student to complete each sheet.

Each of the preceding activities can be modified in several ways. The shape of the grid can change. Different markers can be used to fill in more than one portion of a grid. A part of a block can be shaded to represent a fraction of one percent. In other words, the activities can be adapted to each student's needs.

From the experiences with the meaning of percent expressions, mildly handicapped children can move on to finding a percent of a number. In doing so, we essentially divide the number by one hundred and multiply by

Complete each sentence.

1. $\frac{45}{100} = $ _____ %.
2. $\frac{73}{100} = $ _____ %
3. $\frac{25}{100} = $ _____ %
4. $\frac{30}{100} = $ _____ %
5. $\frac{16}{100} = $ _____ %
6. $\frac{7}{100} = $ _____ %
7. $\frac{3}{100} = $ _____ %
8. $\frac{13}{100} = $ _____ %
9. $\frac{9}{100} = $ _____ %
10. $\frac{5}{100} = $ _____ %

Complete each sentence.

1. $70\% = \frac{}{100}$
2. $35\% = \frac{}{100}$
3. $15\% = \frac{}{100}$
4. $75\% = \frac{}{100}$
5. $80\% = \frac{}{100}$
6. $7\% = \frac{}{100}$
7. $5\% = \frac{}{100}$
8. $50\% = \frac{}{100}$
9. $25\% = \frac{}{100}$
10. $8\% = \frac{}{100}$

FIGURE 7.17 More percent tasks

the number of hundredths named by the percent expression. In practice this is accomplished by multiplying by the fraction name for the percent or the decimal name. These rules should be developed with the students and practice should be provided. Most applications of percent require only this skill, usually in conjunction with another arithmetic operation. Opportunities to apply percent in real contexts are many. None of them should be lost.

SUMMARY

The study of fractions and decimals completes the exposure to the rational-number system for many mildly handicapped children. Experiences designed to highlight the part-whole relationship should precede formal instruction in fractions. The careful and consistent use of manipulatives should predominate the teaching strategies for basic concepts and operations with fractions and decimal fractions. For many of us, the application of that content will be confined to the area of percent. That is why we have concluded this chapter with limited suggestions for teaching the basic concepts of percent.

MEASUREMENT

INTRODUCTION

Learners come to school at the kindergarten level with some measurement concepts that can be assessed by listening. They may talk about a big brother or little sister, how warm it is in the classroom, or who is taller. They are able to verbalize these ideas of measurement because of experiences and the language development that accompanied these experiences.

While the first concept of geometry is relative position, the first concept of measurement is *relative size*. As with geometry, the first task of a teacher is to extend the vocabulary development of the learners in terms of measurement concepts. This usually involves antonyms—tall, short; big, small; long, short; larger, smaller. Again, we deal with real objects such as chairs, books, people, and pieces of wood or rope. First, only two objects are presented; then, three objects are used. Learners should be able to state which object is longer or shorter. They should be able to identify the larger or smaller object by pointing to the appropriate solution. It may be possible to extend these verbal descriptions to include heavier and lighter objects. This vocabulary development foreshadows linear measurement and other general concepts and skills of measuring.

The importance of measurement cannot be de-emphasized since an understanding of measurement will enable learners to function more successfully in society. Shopping for groceries, preparing a meal, and operating a car are a few examples of the social uses of measurement. In the classroom measurement provides the opportunity for an activity-based part of the mathematics curriculum.

GOALS

The major goals of teaching measurement to youngsters are:
1. To develop in the learner a knowledge of how measurement serves to quantify the world in which he or she lives
2. To develop in the learner an understanding of various units of measurement, which will enable the learner to make decisions in daily activities

With goals such as these we need to establish some developmental guidelines. The desired sequence for measurement starts with experiences with arbitrary units and extends to the need for a standard unit of measure. Many experiences serve to foreshadow abstract concepts, and development of measurement is much slower than the development of geometry.

TIME

One of the most used concepts of measurement is time. Many elements of time are developed instinctively and intuitively from birth. Circadian cycles of man and other animals represent one such element. The circadian cycle of man represents the regular metabolic, glandular, and sleep rhythms associated with the twenty-four-hour cycles of the earth's rotation. Placement of events in terms of time are good beginning points for the teaching of measurement. This is a categorization activity. Initially the categories should be as general as the following: day and night; morning and afternoon; winter and summer. Identifying activities of the learner in terms of *when* he or she does the activity is a natural extension of real experiences.

A list similar to the following might be prepared by an instructor and the learners.

Activities for one school day

Watch television	Go to school
Eat breakfast	Brush teeth
Get out of bed	Do chores
Do homework	Feed pets
Get ready for bed	Sleep
Play with friends	

Each activity should be discussed in terms of when each learner does the activity—day or night, morning or afternoon, or both. The A.M. and P.M. labels can also be discussed in terms of *before* noon and *after* noon. This development should continue. Pictures of activities (see Figures 8.1 and 8.2) should be used, and the learners should be asked to tell (state) when the activities occur.

FIGURE 8.1 Beach scene (afternoon)

FIGURE 8.2 Breakfast (morning)

An extension of the categorizing activity is a sequencing activity, as in Figure 8.3, in which the learner is presented with a set of separate pictures and is asked to put them in order to tell a story. This is a first-level problem-solving activity and prerequisite behavior in regard to developing the concept of measuring time. In order to comprehend the order of days and months, the learner must have understanding of the order of events in a sequence. Vocabulary words indicating *before* and *after* should be developed fully.

Another prerequisite activity in preparation for calendar development is the relating of various seasonal activities to winter, spring, summer, or fall. Pictures representing various seasonal activities can be used here; for example, a beach scene for summer, a snow scene for winter, flowers for spring, and falling leaves for fall.

Development of a Calendar

The practice of keeping a daily and/or weekly calendar in the classroom is a good measurement activity as the students learn the sequence of the days of the week. The learners should be able to state and write the names of the days of the week and identify dates on a calendar. A week should be defined as a time period of seven days, and learners should be asked to draw a calendar when given a blank form such as the one in Figure 8.4.

Significant holidays should be defined and placed (named) on the calendar. Birthdays can also be defined and named on the calendar. When looking at their calendars learners should be asked to determine how many days it is until some selected date; for example, until a vacation period or until someone's birthday.

Calendar activities can be personal activities and are thus motivating ones

FIGURE 8.3 Pictures to put in sequence

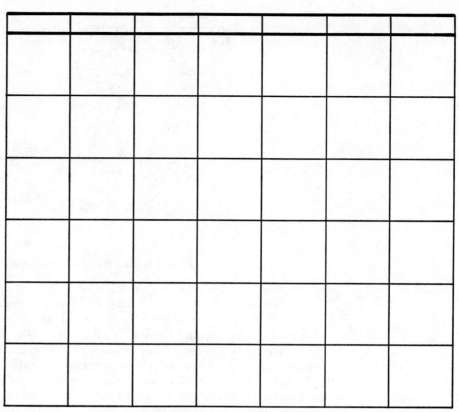

FIGURE 8.4 Form for calendar

for the learners. Calendar activities can be extended to such problems as determining the age of people and events when the current date and relevant information are given. Writing dates in the following forms should also be emphasized: (1) January 1, 1987 or (2) 1/1/87. If you have access to a computer and printer, a teacher utility program can be used to print calendars for the learners.

Time on a Clock

From calendar development, the measure of time is extended to telling time on a clock (a nondigital one). Working with a clock can also be used to demonstrate how a measuring scale should be constructed. The face of a clock should be presented as in Figure 8.5, with only the whole numbers showing—no hands at first. Be sure to use a model for the clock. Learners should also have a model if possible. The hour hand should be added and placed only on the numbers listed. Attention is first called to the clockwise movement of the hour hand, as indicated in Figure 8.6. Add the minute hand and place it toward twelve, as in Figure 8.7. Move the hour hand and ask the learners to state (name) the time. Specify whether the time is *before* noon or *after* noon. We should note at this point that we are outlining a complete development. Not all of the activities are to be taught at the same time or within one grade level.

At an appropriate stage, minute "marks" should be placed on the clock (see Figure 8.8), and a scale should be labeled until the student learns what the marks mean. Place the hour hand in one position and leave it there, or temporarily remove it as the minute hand is moved through each of the sixty

**FIGURE 8.5
Clock face**

**FIGURE 8.6
Hour hand on clock**

5:00 p.m.
Afternoon

8:00 a.m.
Morning before noon

2:00 p.m.
Afternoon

FIGURE 8.7   Minute hand on clock

minutes. Teaching the students how to tell time is teaching a procedure that does not vary; that is, a definite sequence of steps should be used.

1. Look to see where the hour hand is pointing. If it is anywhere from the number to just before the next number, it is read as the first number (see Figure 8.9).
2. Next, look to see where the minute hand is pointing. Count the number of minute marks from the top of the clock, the twelve, as in Figure 8.10.

The learner should be able to *tell* (state), read, and write time to the minute. Seconds should be introduced by counting and watching a second hand on a watch or clock. The relationships of 60 minutes to the hour and 60 seconds to the minute should be identified in the middle school. Careful, sequential development here will assist the learners elsewhere; for example, the development of a ruler. Another foreshadowing or exploratory activity that can be included is that of *estimating* time—10 seconds, 30 seconds, 60 seconds or a minute, and 5 minutes—without and with interruptions.

After the development of the clock in this form and it is clear that the learners understand the concepts of hours and minutes, the digital clock should be used. Under these circumstances we have used time to develop measurement concepts that could not have been completed if a digital clock had been introduced first. Learners should now have a meaning for a digital reading of 7:14 P.M.

A C T I V I T Y

SAMPLE INSTRUCTIONAL ACTIVITIES FOR TIME FOR MIDDLE SCHOOL LEARNERS

1. Show the learners pictures of activities such as those listed after this paragraph. Ask the learner to describe the time of day (morning, noon, afternoon, night) during which each activity takes place.

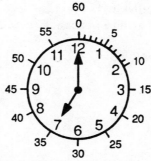

FIGURE 8.8   Minute marks on clock

5: ____   5: ____   5: ____

FIGURE 8.9   Position of hour hand

   a. A child eating breakfast
   b. A family eating dinner
   c. A child getting ready for bed

2. Use a large cardboard clock with movable hands. Set the clock to the following hours and ask the learner to state the time.

   a. 3:00
   b. 1:00
   c. 8:00
   d. 6:00

3. Give the learner a demonstration clock. Ask the learner to set its hands to

   a. one o'clock
   b. six o'clock
   c. eight o'clock
   d. ten o'clock

4. Set a model clock to the following items: 11:41, 12:30, 3:45, 4:40, and 7:35. Ask the learner to write the times as minutes before and after the hours.

5. Use a page from a calendar and ask the learner to answer the following questions:

   a. What day of the week is the fifteenth?
   b. What day of the week is the twenty-fourth?
   c. What is the date of the first Tuesday?
   d. How many Wednesdays are in this month?

_____

This outline of time demonstrates the slower development of measurement concepts. For example, this development covers at least four years to produce a good understanding on the part of the learners.

TEMPERATURE

Hearing temperature readings and comparisons is a common occurrence in an individual's life. Learners often hear "It's too cold to go swimming" or "It's too warm to wear that heavy coat." Learners are aware of temperature in that

6:07 p.m.

Evening

FIGURE 8.10   6:07

it restricts or expands the types of activities they can do and the type of clothing they wear. For this reason, the first kind of tasks on temperature should deal with temperature in relation to seasons, events, and clothing. The activities from the preceding section on time with respect to the seasons can include information on temperature. Seasonal activities such as playing baseball, ice skating, fishing, sunbathing, and going to an amusement park can be related to seasons and to temperature ranges. Pictures from different parts of the world can also be used to discuss temperature. Manner of dress and clothing also reflect temperature ranges.

Using the preceding information as introductory activities, the instructor can proceed to the next step, which involves placing a thermometer with a designated temperature reading with each picture. We should note at this point that both the metric and English systems of measurement should be used, but they should be used independently. Any conversions should be left to the learners to develop for themselves.

Samples

Baseball game	70° Fahrenheit
Ice skating party	28° Fahrenheit
Room temperature	22° Celsius
Ice water	−2° Celsius

Development of thermometers should follow scale-development procedures. If possible, uncalibrated bulbs should be fastened to tongue depressors and placed in ice water for five minutes. The level of alcohol or mercury in the bulbs should be labeled as 0° Celsius and 32° Fahrenheit for freezing point. Remove the bulbs, let them warm, and then place them in boiling water for two to four minutes. Mark the new level and label it as 100° Celsius and 212° Fahrenheit for boiling point. The Celsius scale can be divided evenly into 10 divisions from the 0 to the 100 and labeled accordingly. The Fahrenheit scale requires 18 divisions from 32 to 212. After the learners make some initial readings with the homemade thermometers, they can be given standard thermometers to use. Freezing and boiling points should be verified with these thermometers. Body temperature should be indicated for each scale and the students should learn what levels the temperature readings indicate what is usually uncomfortable; for example, less than 30° and greater than 90° Fahrenheit.

Models of thermometers, such as the one in Figure 8.11, can be used to indicate various levels as the learners identify freezing point, boiling point, and appropriate temperatures.

The importance of knowing about temperatures can be identified by using real-life examples. Since learners know that during the winter season temperatures go below freezing (at least in the northern parts of the United States) and cars with radiators full of water must be protected because water freezes at 32° F, they should learn that antifreeze solutions lower the freezing point. Label reading can be of help here, and ratio and proportions can be reviewed. (Note: This procedure identifies the placement of the scale-development concept after ratio and proportion has been developed.) Oven temperatures for cooking items such as a pizza can also be explored. Other supplementary activities involve reading dials and other types of scales. Knowing what the readings should be gives positive reinforcement to the learners and provides them with a solid background in measurement. Activities such as these are often neglected in regular mathematics curriculum, and mathematical gaps are created. Without experiences in scale development and reading, the learners have a "built-in" disability. Perhaps many of the learning disabili-

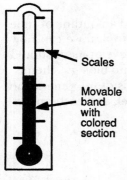

**FIGURE 8.11
Thermometer model**

Scales

Movable band with colored section

FIGURE 8.12   Sorting by temperature and clothing

ties are due to faulty developmental activities or the lack of them. Determining temperature differences between above-zero readings and below-zero readings and estimation of temperatures should be delayed until the learner has had experience with the operations of integers and is in the formal stage of intellectual development.

A C T I V I T Y

SAMPLE INSTRUCTIONAL ACTIVITIES

1. Show the learner sets of pictures such as the ones illustrated in Figure 8.12. Ask the learner to tell which picture in each set does not belong with the other pictures.

2. Dress a paper doll for three distinct temperature conditions. Suggested modes of dress are bathing suit, snow suit, and sweater. Ask the learner to place a marker on a picture thermometer to show when each item of apparel would be appropriately worn.

3. Show the learners a representation of a Fahrenheit scale and a representation of a Celsius scale. Ask the learners to locate 48° on the Fahrenheit scale; 20° on the Celsius scale.

4. Write the following temperatures on cards or sheets of paper: 25° C, 25° F. Ask the learner to tell how each temperature would feel, what he or she might wear, and what he or she might do.

With the idea of relative size as a beginning concept and an outline of time and temperature to illustrate the scope of measurement topics, we can now discuss a sequence for measurement. The measurement development should go from kindergarten through eighth grade and extend into biology, chemistry, and physics at the senior high school level. Within this development the spiral curriculum idea should be present at each grade level, and each opportunity to enhance problem solving should be used.

MEASUREMENT OUTLINE

Relative size

   Ordering by size
   Size patterns

Time

   Sequencing events
   Clock development
   Calendar development

Temperature

   Sequencing events
   Thermometer development

Linear measure
    Arbitrary units
    Standard units of linear measure
Area
    Polygons
    Surface area
Volume
Weight

More specifically and in conjunction with learning theory regarding readiness, development, and reinforcement the following list of specific activities has been developed.

ACTIVITY

GRADE-LEVEL ACTIVITIES

K, 1

Sorting objects of various lengths and weights
Comparing objects with respect to length
Using objects approximately a decimeter long or a foot long
Working with size and shape
Making arbitrary units such as "foot" patterns

2, 3

Using meterstick or yardstick as a number line
Developing standard for length
Working with inequalities
Reading temperature

4, 5, 6

At this stage measurement becomes operational, and the learners should become engaged in the following types of activities:

Discovering the decimal nature of the metric system
Extending the use of prefixes and establishing their relationships
Approximating lengths and checking by measuring
Performing basic operations with measuring instruments
Exploring volume and capacity—readiness activities
Weighing objects
Timing events

7, 8, 9

Measurement can be refined and reinforced at this stage of development; therefore, the learners should be doing the following:

Using formulas
Developing the relationships between the standard units of measure
Seeing the mathematical implications and properties of measurement
Working with approximations

These activities sample the areas and serve as starting points for the development of more complete lists. It should be noted that if prerequisites from one level to the next have not been acquired, then the learners should be exposed to those prerequisites and achieve a level of mastery before going ahead.

LINEAR MEASURE

As learners are describing the characteristics of three-dimensional objects we can begin to ask how long the edges of a particular object are; thus, we introduce linear measure. Some familiar, nonstandard unit of measure should be the first measuring device used. A chain of paper clips of the same size is just one example of such a device. Patterns made from tracings of a learner's hand or foot are other examples. The first act of measuring involves counting how many units can be placed beside the object to be measured. Note that these arbitrary, nonstandard units of measure have no divisions (marks) other than the whole object; therefore, the answers should be whole-number answers—so many paper clips long or so many hands long. This approach should be continued as standard units of length are introduced. A piece of cardboard or oaktag with no markings should be cut to a decimeter length or a foot length and used as a first unit of measure. Some individuals would suggest a 10-inch ruler instead of a 12-inch (foot) ruler for more precise development of units. We should also note that both the metric system and the conventional system of inch, foot, yard, and so forth should be developed with the learners. They should be developed independently with no formal conversions between systems being taught before high school. The developmental process should be free exploration, guessing or estimating, observation through experimentation, and verification.

Starting with an arbitrary unit, an instructor must redesign the activities to fit into a particular situation or grade level.

ACTIVITY

SAMPLE INSTRUCTIONAL ACTIVITIES

1. Ask the learner to measure the height of a chair using his or her hand as a measuring tool. Next, ask the learner to use pencils to measure the bulletin board. Following each measure, ask the learner to state the measure in terms of the tool used.
2. On a desk, place pieces of yarn that vary in length. Give the learner a pool of one-inch cubic blocks and ask him or her to construct a row of blocks the same length as each piece of yarn.
3. Place a small box on a table. Show the learner an eraser, a book, and a yardstick. Ask the learner to choose the item that will fit in the box. Have the learner place the item in the box to verify the correctness of his or her choice.

Using nonstandard units as an introduction to the measurement process, we duplicate what individuals did hundreds of years ago. The first units of measure were very arbitrary and were devised to serve a particular need. As long as individuals remained within their own communities, little need for change occurred; however, as travel became possible and important, the need for comparing the arbitrary units of various peoples became evident.

Selecting a "Best" Unit of Measure

Some of the considerations in determining a "best" unit of measure are:

1. Everyone using the unit should get the same whole-number answer. Would this always be true for an $8\frac{1}{2}'' \times 11''$ sheet of paper? Only if the paper is made into a square shape would it be true (e.g., $8\frac{1}{2} \times 8\frac{1}{2}''$). Some individuals might use the shorter side to measure if it is rectangular in shape.

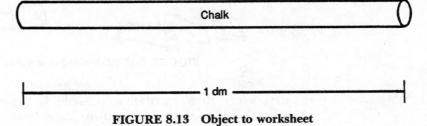

FIGURE 8.13   Object to worksheet

2. The unit should be easily used and reproduced.
3. The unit should be the most accurate—closest to a true measure.

Arbitrary units meet one or two of the criteria, but none satisfies all criteria; therefore, as in ancient times, there is a need for a standard.

Introduction to a Standard for Length

Working with the metric system and our conventional system of measurement we have two opportunities to develop the measurement concepts and skills. Let's begin with the metric system. Give the learners a decimeter length of cardboard and ask them to measure a "standard" sheet of paper ($8\frac{1}{2}''\times 11''$) and use only whole-number answers (round off). Guide the learners to obtain a width of 2 decimeters and a length of 3 decimeters. Measure other objects in the room and record answers. (Note the laboratory process of observation and recording of answers.) Make a list of objects in the room that are a decimeter in length. Provide a worksheet for the learners that contains pictures of real objects that have lengths that can be expressed as whole decimeters, *not* fraction answers (as $1\frac{1}{2}$ (1.5) or $2\frac{1}{4}$ (2.25) decimeters).

Developing Measurement Worksheets

As we move from the object mode to the pictorial mode for our presenting of information to the learner, we need to make the step as small as possible. The construction of worksheets for measurement activities will illustrate this suggestion. The pictures of a first measurement worksheet should be pictures of actual objects in actual size. For example, as shown in Figure 8.13, a piece of chalk and its picture should be in true size to enable the learners to place the object on the picture for comparison purposes.

The second step, as depicted in Figure 8.14, involves line segments that are the same length as selected objects. Using this approach enables an instructor to go from an object to its symbol to a measuring device.

A parallel development of the foot unit should now be introduced to reinforce the development of a standard for length and to extend the development.

A C T I V I T Y

SAMPLE INSTRUCTIONAL ACTIVITIES

1. Select some common material such as paper, paper tape, string, or nonstretchable ribbon. Prepare a representation of a meter length by using

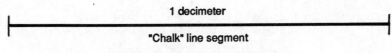

FIGURE 8.14   Line segment

FIGURE 8.15 Developing a meter length

ten decimeters. Be as accurate as possible. If paper tape or ribbon is used, place cardboard tabs over the ends to make them more stable, as shown in Figure 8.15. Such material is suggested because of its flexibility quality. It will be used to measure curved paths and surfaces.

2. After constructing a meter length, ask the learners to become familiar with this length by estimating a horizontal, a vertical, and two slanting distances that appear to be a meter in length. Check the estimates by using the constructed meter length. Practice until they can make a good estimate without a measuring device.

3. Next estimate, measure, and record the measurements of the following (remember to use only whole numbers for answers).

Width of door	——meter(s)
Height of door	——meter(s)
Width of room	——meter(s)
Length of room	——meter(s)
Length of hallway	——meter(s)

4. Continue the activities by determining a metric (meter) pace. That is, walking naturally, how many meters do you and the learners cover in one step, two steps, or three steps? The average is two steps per one meter or three steps per two meters, depending on height and stride. To begin, place a constructed meter length on the floor and determine a metric pace.

My meter pace is ——— steps per ——— meter(s).

5. Once you have determined your pace, look at the following table and study the distances covered in various numbers of steps.

Distance

Steps	One step/ One meter	Two steps/ One meter	Three steps/ Two meters
1	1 m		
2	2 m	1 m	
3	3 m		2 m
4	4 m	2 m	
5	5 m		
6	6 m	3 m	4 m
7	7 m		
8	8 m	4 m	
9	9 m		6 m
10	10 m	5 m	
11	11 m		
12	12 m	6 m	8 m
13	13 m		
14	14 m	7 m	
15	15 m		10 m

It may be necessary for you to develop a fourth column if you do not fit into the ones given. Note that only whole-number answers have been used. The use of fractional numbers would be discouraging, so record the distance to the nearest whole number.

The development of the yard parallels the meter development, and the answers are almost the same, since a meter is a little more than a yard.

Standard Units of Linear Measure

After we have developed a longer unit of measure we should continue to work toward more accuracy by developing a smaller unit of measure; for example, the centimeter from the decimeter (10 centimeters = 1 decimeter) and the inch from the foot (12 inches = 1 foot). At this point the measuring device should have only the centimeter and decimeter divisions or only the inch and foot divisions. Learners should be asked to record the measurements to the nearest whole-number mark. Initially each centimeter division should be labeled on the measuring instrument. Ask learners to make a list of objects that are one centimeter in length or width, such as the width of a fingernail on your little finger. Then ask the learners to make a list of objects that are one

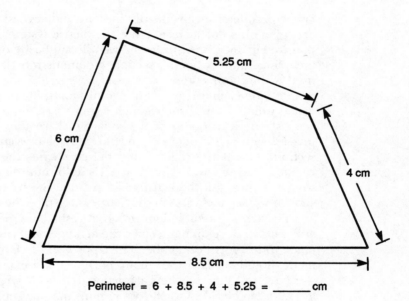

Perimeter = 6 + 8.5 + 4 + 5.25 = _____ cm

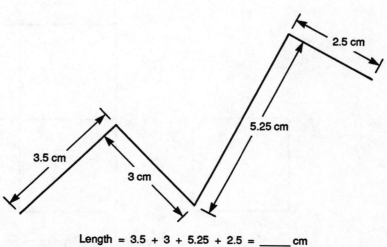

Length = 3.5 + 3 + 5.25 + 2.5 = _____ cm

FIGURE 8.16   Perimeter and length

inch in length or width, such as the width of a thumb. The "standard" sheet of paper ($8\frac{1}{2}'' \times 11''$) now has dimensions of 22 cm by 28 cm, and the learners can be led to *discover* that the distance around (*perimeter* of) the sheet of paper is approximately one meter. With a flexible measuring device (a tape), the concepts of perimeter and circumference can be developed within a meaningful context. Figure 8.16 shows the geometry relationship: the perimeter of a polygon is the sum of the lengths of each side of the polygon; the length of a broken-line path is the sum of the lengths of each line segment of the path. The distance around a circle is called the *circumference* and should be measured with a flexibile measuring device. As skill develops, smaller units of measure should be introduced.

As with all measurement, after the learners understand the concept, each measurement activity should involve the following steps: (1) determine what is to be measured; (2) select an appropriate measuring tool; (3) estimate what the answers are for the object to be measured, and record this estimate; (4) use the tool and measure; (5) record the appropriate answer with its label.

AREA

While perimeter means the distance around, *area* means the amount of space occupied on a flat surface. Since the metric system of measurement was emphasized in linear measurement, we will emphasize the conventional system of inch, foot, and yard in this section to demonstrate that similar approaches are used within both systems.

Provide each learner with several square-inch regions and polygons into which will fit a set of square-inch regions (see Figure 8.17).

Ask the learners to fit the square-inch regions into the polygons like a puzzle, with sides touching and no gaps or overlap (or cover the enclosed section). Count the regions and define *area* as the number of square-inch regions used to cover the surface. Using 6 square-inch regions, an area of 6 square inches, ask the learners to arrange the 6 regions in other shapes to show 6 square inches. Ask the learners to trace their figures.

The arrangement in Figure 8.18 foreshadows the concepts of surface area and pattern development, since the arrangement can be cut out, folded, and made in the form of a cube. This approach of "filling in" for area development should be used with many polygons and extended so that the learner will have to estimate when a fractional answer is required. On a geoboard, learners can be asked to enclose regions that have an area of 1 square unit, 2

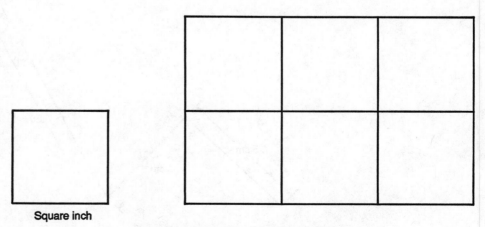

Square inch

FIGURE 8.17   Square inches for area

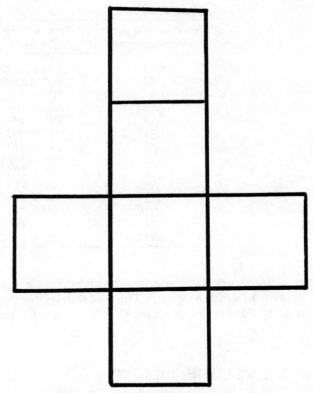

FIGURE 8.18   An area of 6 square inches

square units, 3 square units, . . . ; then, 1/2 of a square unit, 2/2 of a square unit, 3/2 of a square unit, ... This suggestion serves to place this area development after an understanding of fractions. This is also an appropriate place to introduce area-formula development for a right triangle. Area formulas for the square and rectangle should be developed after multiplication of whole numbers.

Wrapping a package can be used to demonstrate surface area. Lay a box on a piece of paper and trace all sides by rolling the box until you have a section for each surface. Learners will experience the concept of allowing extra paper for overlap and attaching one surface to another. Estimation plays a role here.

The units for areas should be built from smallest to largest—for example, square inch, square foot, and square yard.

Applications should be introduced with this topic by working on such exercises and problems as: How much floor space (area) does a given room or building have? What kind of regions fit together to cover a surface? How many square yards of carpeting will be needed to cover the hallway? How many square yards of material are needed to make a skirt or shirt? Some interesting problem situations might evolve. How big is a parachute? How much glass does it take to construct a forty-liter aquarium?

In the metric system, a good unit with which to begin is a square decimeter, since its size is between a square inch and a square foot. Bathroom tiles have this area, so models are readily available. Development with this unit should be as complete as with the square inch and square foot and should be extended to the square meter.

An application and problem area for square units in the metric system exists with international paper sizes—the A series with A0 equal to one square meter. The one square meter is for the area of a rectangle that has dimensions such that if the rectangular region is cut in half across the width, the next size

TABLE 8.1

International paper size designation	Trimmed sizes (in mm)	Some possible uses
2A0	1189 × 1682	Posters
A0	841 × 1189	Maps
A1	594 × 841	Gift wrap
A2	420 × 594	Newspaper (sing. sheet)
A3	297 × 420	Tabloid newspapers
A4	210 × 297	Letter size
A5	148 × 210	Book size
A6	105 × 148	Postcard size
A7	74 × 105	Index cards
A8	52 × 74	Credit cards
A9	37 × 52	Admission tickets
A10	26 × 37	Postage stamps
A11	18 × 26	Labels
A12	13 × 18	Stickers

paper, A1, is produced with very little waste. A0 and A1 have the same ratio of width to length. The problem is to determine the dimensions of the various sizes in the A-series. (See Table 8.1 for the solution.)

VOLUME

Characteristics of solid figures and models of these figures contain the prerequisite material for volume; however, this topic should be delayed until an understanding of area has been achieved. The development of surface area also contributes to the development of volume. Surface area is related to the amount of material it takes to construct a container; *volume* concerns the *size* of the container—how much is inside. Cubic units are used to measure volume.

Intuitive ideas of volume can be obtained by exposing the learners to different sized containers in the form of boxes, cans, and bottles. The vocabulary development of *more* or *less* can be extended to such facts as "A liter is a little more than a quart." Relationships among small containers and larger containers can be established by pouring water or some other material from one container to another.

An interesting relationship exists for a cubic decimeter. The volume of this cube is 1 cubic decimeter or 1000 cubic centimeters. The amount of water that you can pour into a container one cubic decimeter in size is 1000 milliliters or one liter, and the approximate mass of this one liter of water is one kilogram.

Formula development for volume should be delayed well into junior high school and should follow three-factor multiplications, such as (3 × 9) × 8. If this type of multiplication has been neglected, then the development of volume formulas requires an extra step.

Stacking blocks, centimeter cubes, or inch cubes to produce models of geometric solids is an excellent way to develop the relationships from sets of points (the objects, cubes) to sets of numbers (how many cubes are used to construct the figures or what number represents the volume). This strategy also leads learners to conservation of volume as they can see that a stack of cubes one block by four blocks has the same number of cubes as two blocks by two blocks (see Figure 8.19).

Volume involving cubic measures, whether in the metric system or our conventional system, is somewhat structured, but liquid volume in the conventional system is somewhat unstructured because of the many units involved.

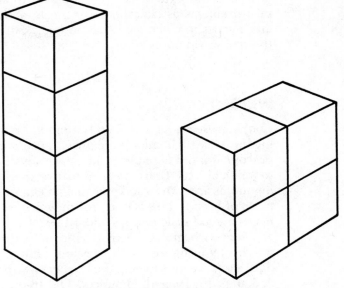

FIGURE 8.19   Volume of 4 cubes

SAMPLE INSTRUCTIONAL ACTIVITY

1. Using the relative-size concept, collect nonstandard containers: bottles, jars, coffee cans, buckets. Using a measuring cup with a one-cup mark, demonstrate and then ask the learners to do the same by pouring two cups of water into one of the containers. Using tape or markers, the learners should identify, on the side of the container, where the level of the water is and write that 2 cups = 1 pint. Next, pour the water from one container to another, labeling as you proceed. Continue this process until all of your nonstandard containers have the one-pint label. Show the learners a pint container of milk. As we have suggested with other measures, we need to extend and break down our beginning unit. After learners have experience with this unit, the half-pint should be introduced since we find such containers in use.

2. Using the pint as a standard, we should now pour 2 pints in our containers and label 2 pints = 1 quart to introduce the quart. We should repeat this procedure using 4 cups to produce the 1 quart. After such a development we should introduce the quart of milk. Learners should be asked to make lists of materials that are measured in cups, pints, and quarts— for example, milk for cereal, yogurt, and orange juice, respectively.

3. The development continues with large containers. Using the quart as a standard, pour 4 quarts into the large containers and label 4 quarts = 1 gallon to introduce the gallon. Repeat using 8 pints. Repeat using 16 cups. A gallon of milk or paint can be displayed for the learners. Use as many real life examples as possible to assist the learners.

Since our measure development is at least a middle school activity, some possible worksheets include the following:

1. Making lists or matching appropriate units of liquid volume to measure selected materials.
2. Relationships of units—for example, _____ pints = _____ gallons.
3. Word exercises and problems—for example, Janet got 1 gallon of milk from the refrigerator and poured 1 cup to drink. How much milk was left?

We could extend our development to smaller containers and use such measurements as tablespoons and teaspoons within the domain of cooking and recipes to make the situation real. Many mathematics programs neglect the real-world application.

WEIGHT

Cooking also serves as a good beginning for using *weight,* since recipes involving meat use such units of measure. A mathematics program would be lacking in scope if it did not at least introduce *ounces* and *pounds* and the use of scales to weigh objects. Learners need many experiences to develop an understanding in this area. Did you know that an egg weighs about two ounces, as does a piece of bread? This is one area in which we need equipment. Each teacher needs a good scale and a standard set of weights.

A good beginning activity is to assist learners to make their own set of weights by using sandwich bags and sand or gravel. Use the scale and standard set of weights to help the learners weigh and label bags of material for 1, 2, 3, 4, 5, 10, 16, and 20 ounces. The 16-ounce bags should also be labeled 1 pound. The learners should use these "developed" standards to weigh other objects on a balance-beam scale.

We should also capitalize on the current trend in body building by discussing weights used in lifting activities and the various size weights; for example,

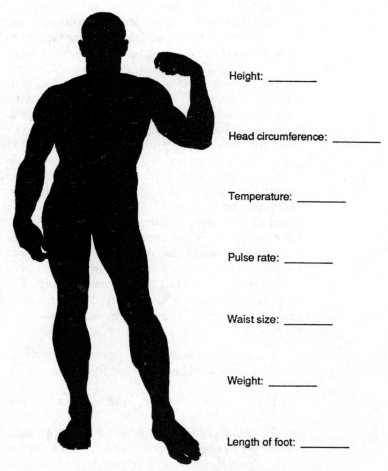

Height: _____

Head circumference: _____

Temperature: _____

Pulse rate: _____

Waist size: _____

Weight: _____

Length of foot: _____

FIGURE 8.20   Measurement of the body

5 lbs, 10 lbs, 15 lbs. Addition and subtraction involving units can now be placed in real situations.

MEASURING PROJECTS

An interesting project is to ask the learners to describe themselves using measurement. The amount and degree of difficulty depends on where the learners are in the measurement development; that is, how many measurement concepts do they understand? The following outline of questions represents a wide range of possibilities. For some questions several answers are requested (see Figure 8.20).

Question	*Answer*
How tall am I?	_____ inches
	_____ feet _____ inches
	_____ centimeters
	_____ decimeters _____ centimeters
	_____ meters _____ decimeters _____ centimeters
How far is it around my head—my cap size?	_____ inches
	_____ centimeters
How long is my right arm?	_____ inches
	_____ centimeters
How far is it around my left wrist—my watch size?	_____ inches
	_____ centimeters
How far is it around my waist—my belt size?	_____ inches
	_____ centimeters
What is the distance around my knee?	_____ inches
	_____ centimeters
How long is my right foot?	_____ inches
	_____ centimeters
How big is my smile?	_____ centimeters
How far is it from fingertip to fingertip if I stretch my arms to the sides of my body?	_____ inches
	_____ centimeters
How much do I weigh?	_____ pounds
	_____ pounds _____ ounces
	_____ kilograms
How much does my shoe weigh?	_____ pounds
	_____ pounds _____ ounces
What is my pulse rate?	_____ beats per 10 seconds
	_____ beats per minute

What is my temperature? _____° F
_____° C

How many minutes did I
sleep last night? _____minutes
_____hours_____minutes

How old am I? _____years_____months_____days
_____hours
_____years
_____total months
_____total days
_____total hours

Some questions that require reading for information:

How many pints of blood
do I have? _____pints

What is my blood
pressure? ____/____

What do the blood
pressure readings
mean?

Considering my height,
how much should I
weigh? _____pounds

What are calories?

How many calories are
in the foods that I ate
yesterday? _____calories

How much water does my
body displace in a
pool? ____

An instructor may extend this project with questions appropriate to a given age and grade level. Learners may pose additional questions.

Other projects may be developed around the Olympics that are held every four years. What are the records in various events? How have the records changed in the last sixteen years? Graphing the events and changes can provide an opportunity to relate numbers of measurement to numbers of arithmetic.

Planning a camping trip can also be a mathematical experience, involving several units of measure. Each learner may be given a task that matches his or her interests and strengths. We need to take every opportunity to enhance measurement, since this area is often neglected in mathematics.

SUMMARY AND CONCLUSIONS

In this chapter we have attempted to outline a developmental approach to teaching measurement (time, temperature, linear measure, area, volume, and weight) to mildly handicapped learners. This set of materials represents an extensive development for measurement; it is our desire that teachers extend this development.

CHAPTER NINE

PROBLEM SOLVING: WORD PROBLEMS

Problem solving is integral to any curriculum or topic that is presented to children with handicaps to learning and/or achievement. For the most part, educators have allowed rote computation to dominate curricula and instructional experiences in programs for all children, although there is increasing evidence that some changes are taking place (Resnick and Resnick 1985) and that more changes are being sought (NCTM 1980). Problem solving has been relegated to a secondary role in programs for the mildly handicapped because professionals and laypersons have generally been of the opinion that one cannot solve problems unless one has acquired certain computational proficiencies and demonstrated competence in related skills such as reading.

Our perspective differs. We are convinced that meaningful quantitative problem solving can take place before one has acquired extensive computational proficiency and before one reads. Mathematics for the mildly handicapped should be concept-based; problem solving is essential for qualitative programming. In effect, mathematics has to do more for the mildly handicapped child than teach mathematics. The stimulus properties of mathematics need to be used to enhance language development, facilitate cognitive processing, provide applications for reading, and formulate the basis for long-term capabilities in decision making and argument. If mathematics cannot do more for the mildly handicapped child than to teach the child mathematics, programs in mathematics are not as fully conceptualized as they ought to be (Cawley 1984a). This chapter and the next one will provide a sufficient number of examples such that a compelling case for a greater problem-solving emphasis will be demonstrated.

The act of solving problems is a complex one involving many components. Some of these components are the following: (1) the metacognitive acts of knowing that there is a problem, that planning and monitoring are required to solve the problem, and that evaluation and critique of the responses are necessary; (2) the cognitive acts of thinking, reasoning, and the search for procedures that will produce one or more desired responses; and (3) the appropriate use of skills to obtain the desired response.

For purposes of this text, problem solving will be viewed as a process that individuals or groups perform to (1) generate acceptable responses that are not automatically generated, (2) clarify an ambiguity or a discrepancy, (3) determine an unknown, (4) provide alternatives in given situations, or (5) generate situations or instances that do not presently exist.

We view all problem solving as complex. Fundamental to this is the line of thinking in Shaw's (1981) differentiation among problems, examples, and exercises. Shaw indicates that problem solving takes place at three levels. The first is the actual confrontation with a problem. Here the child recognizes that a problem exists and also recognizes that he or she has no immediate or decisive solution at hand. The second stage is the exercise stage. At this stage the child recognizes the item as one with which he or she has had experience and then draws on that experience to develop a response. The child is deliberate and methodical, but by no means absolutely certain of the outcome. The third stage, the example stage, is that level when the child can glance at the item, recognize immediately what it involves, select the proper steps, and know full well that the response will be correct. Mildly handicapped children seldom reach the third stage, which is measured by a combination of speed and accuracy across problems of varying types. Our goal is to assist them to reach this third stage. For many children, the items provided under the heading of problem solving are merely exercises. For a smaller group of others, they are truly problems, but they are treated as exercises by teachers and by material developers in that no instruction in problem solving is provided, and each child is expected to produce the expected response.

THREE TYPES OF PROBLEMS

School programs are dominated by three types of problems: word problems, knowledge-based problems, and decision-making or argument problems.

Word Problems

Word problems are those items in which the words and their structures create problems. Word problems should not be computational tasks embedded in words. Analysis and interpretation of information must form the bases for making selections and decisions. Most word problems in school texts can be solved by selecting and utilizing the operation stipulated by cue words (e.g., *left, remaining,* and *gave away* to signal subtraction). Schoenfeld (1982) notes that 97 percent of the word problems in one elementary series would yield the expected response by use of the cue-word method.

Programs of the cue-word type appear detrimental to mildly handicapped children because they encourage the individual to bypass the information set, to ignore the processes of analyses and interpretation, and to circumvent the metacognitive sequence. In effect, they do not encourage problem solving.

Subject Area Applications

A second form of problem solving in school curricula focuses on applications in areas such as science, recreation, and vocational education. A major difference between problems of this type and the typical word problem is in the need for specific knowledge. Another difference rests in the number of steps or actions needed to obtain the data and to process it to a successful solution.

A football team has the ball on the opponents' 8-yard line. The opponents commit a personal foul and are penalized. If they are penalized one-half the distance to the goal line each time they commit a personal foul, how many times can they foul before the team with the ball gets a touchdown?

Solving problems such as the preceding requires knowledge that is specific to a given topic. A youngster might fail to solve the problem due to a lack of knowledge about the rules of football, which do not allow a touchdown on a penalty; or the youngster might not be knowledgeable of the principle of density, which indicates that between any two fractions there is space for another. In effect, there will always be half the distance.

Decisions and Arguments

A third and somewhat less common mathematical problem-solving activity stresses data collection and analysis to assist in decision making or in the development of arguments. Examples might focus on a determination of the type of automobile preferred by boys and girls in a given class, grade, or school; or on an analysis of the costs of three types of hazardous waste removal. Automobile preferences would be determined when the data have been collected and analyzed and the results made available. Cost factors could easily be negated if the hazardous waste were to travel through a certain neighborhood. Decisions and arguments may or may not be helped by the available data. There are some who would not want hazardous waste to travel through their neighborhoods no matter how small the cost.

The remainder of this chapter will focus on word problems. Knowledge-based problems and problems leading to decision making will be focused on in the next chapter.

SOLVING PROBLEMS

Let us proceed under the assumption that success in mathematical problem solving is contingent on (1) numerous and long-term experiences that maintain the integrity of the problem-solving processes, (2) the availability to the child of the set of knowledge appropriate for the problem-solving tasks, (3) possession of a system to bypass the traditional prerequisites such as reading or computation, (4) an understanding by teacher and child of the characteristics that differentiate problems, and (5) the ability to monitor or guide the self as the recipient of instruction and as the independent performer during problem solving.

For purposes of this chapter, mathematical problem solving is defined as the interpretation of information and the analyses of data to arrive at a single acceptable response or to provide the bases for one or more arguable alternatives. To these ends, word problems provide the initial interaction with problem solving in school curricula. Unfortunately, word problems have been incorporated into the school programs to provide follow-up activities to computation. Word problems have not been given their proper role as stimulants to thinking, to involving children in quantitatively driven activities, or to developing sequences of behaviors that integrate metacognitive processes with special needs in language and other topics (Cawley 1984a).

Experiences

Experiences in mathematical problem solving and in the integrating of metacognitive processes into the problem-solving activity need to be planned

Display 1

A. Categorization/Indefinite Quantifier/Addition

When the children arrived at the Doodlesville barnyard, they saw SOME

> (2) sheep
> (3) pigs
> (1) cows
> (2) chickens

At the Burgertown barnyard, the children saw SOME

> (3) pigs
> (4) cows
> (5) chickens
> (2) sheep

At the Zippyville barnyard, the children saw SOME

> (1) cow and (3) chickens
> (2) chickens and (2) sheep
> (3) sheep and (1) pig
> (4) pigs and (1) chicken

How many farm animals were in all the barnyards together?

> 9
> 11
> 10
> 9

FIGURE 9.1   Level 1 problem solving

and implemented throughout the spectrum of schooling for the mildly handicapped. We believe that problem solving and metacognition are companions. Although there is a degree of uncertainty as to which comes first, it is our conclusion that programs can be developed such that each will play a role.

Figures 9.1 and 9.2 display examples of verbal problem-solving activities, hereafter referred to as VPS1 and VPS2, from *Project MATH* (Cawley et al. 1974, 1976). In the main, these materials were developed to provide young handicapped children with opportunities to participate in problem-solving activities long before they had acquired many of the reading and computational prerequisites common to most programs. Each level contains

- Teacher guide: scripts
- Story mats: large picture scenes
- Object cards: small picture cards

The teacher reads or ad-libs the script. The teacher or the children perform the actions required by the script. These actions and the interactions among teacher and learners comprise problem solving. Note that problem solving in these instances is synonymous with information processing, in that children make decisions based on information, the language of the information, and the manner in which the language is structured.

Notice that the picture mats in VPS1 are all the same: the three barnyard scenes are alike, whereas the small object cards vary. Thus, the problems are created within the object position. This allows the child to function with a minimum amount of new knowledge and stimuli at any one time. Figure 9.1 shows three displays. The language makes use of *indefinite quantifiers*—terms such as *some, many,* and *group* that take the place of a number. The use of the indefinite quantifier forces the youngster to incorporate the pictures into the problem-solving process, because that is the only way the cardinal property can be determined. Should the teacher state the number, the child is not likely to use the pictures. Using both objects or pictures and stating the number at the same time has been one of the major errors in attempts to incorporate pictures into problem solving. The strategy described in this example heightens the attention of the child to relevant information, stimulates the planning process to seek out the appropriate information in a problem, and allows the child to verify his or her response with the information in full view.

The illustrations in Figures 9.1 and 9.2 contain blocks and ovals. The block is referred to as an *option block*. It informs the teacher of the number of object cards for a particular item. Each row through the option block is followed as illustrated. Thus, a number of children can be presented with different sets of information from a single display. Observe the option block in Display 1. Note the use of parentheses. This indicates to the teacher that he or she is not to say the number because the indefinite quantifier *some* is used. The oval contains the answer for a given row. This is important when the teacher has four or five sets of figures to track and there exists the possibility that he or she might lose track of the correct answer.

Notice that the number of boxes in the barnyard is five. Thus, "numberness" is controlled in that the number of object cards is limited by the number of boxes. This was done to limit computational complexity so that the information-processing aspects of the problems could be accentuated.

Figure 9.2 displays a set of story mats and object cards from VPS2. As can be seen, there are four different workers: thin grocery clerk, thin delivery boy, chubby grocery clerk, and chubby delivery boy. VPS2 varies both the

1. How many	apples bananas green fruit yellow fruit pieces of fruit	do the grocers have?	5 14 8 11 19
How many	apples bananas green fruit yellow fruit pieces of fruit	do the delivery boys have?	10 5 10 5 15
How many	apples bananas green fruit yellow fruit pieces of fruit	do the workers have altogether?	15 19 18 16 34
How many	apples bananas green fruit yellow fruit pieces of fruit	do the skinny people have?	7 10 9 8 17
How many	apples bananas green fruit yellow fruit pieces of fruit	do the fat people have?	8 9 9 8 17

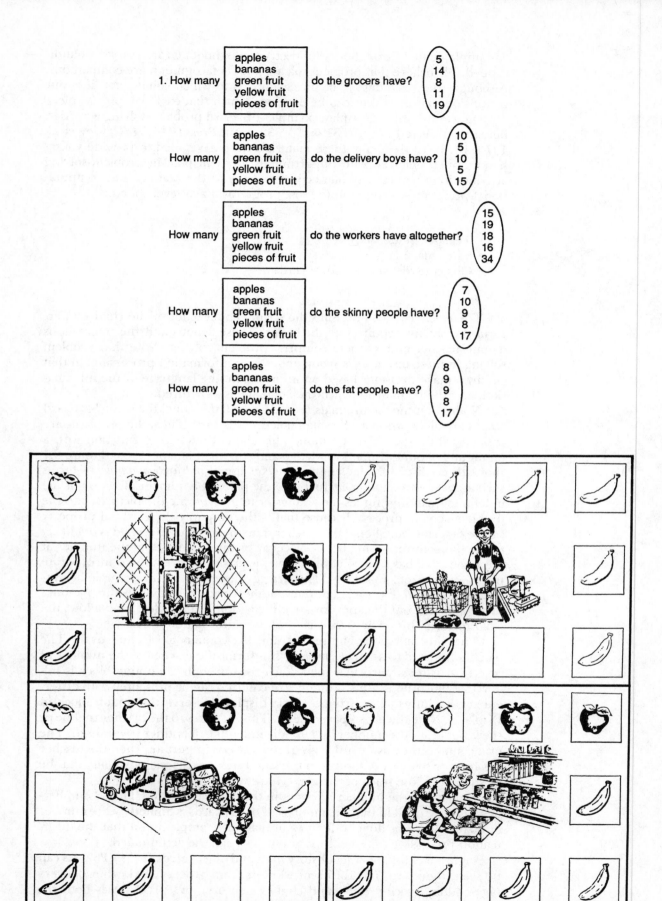

FIGURE 9.2   Level 2 problem solving

subject and the object in the information set. The teacher can ask about any worker or any combination of workers. VPS2 also contains a greater variety of object cards; there is the potential to use up to forty cards in any one setting. The object cards in Figure 9.2 include bananas and apples that are yellow or green. The actual materials are multicolored.

Introductory activities for each level begin with one card and build to 3 cards in VPS1 and 4 cards in VPS2. Problems range from simple to complex within each level. Types of information processing (e.g., extraneous information) are built into each level.

The materials illustrated in Figures 9.1 and 9.2 show quite clearly that problem-solving activities need not be limited to paper-and-pencil worksheets and that a highly interactive and multimodal approach can be developed. Both the teachers and the children become active participants in the activities, thereby creating the potential to deal with Torgenson's (1982) concern for the passive nature of the learning disabled.

Knowledge

Knowledge is essential to problem solving. In some instances, knowledge might refer to general familiarity with the context in which the problems are being presented. In other instances knowledge might refer to explicit rules or procedures that must be followed to accomplish the task.

The problem-solving characteristics of the mildly handicapped child should not be misinterpreted. For this reason, as illustrated in Figures 9.1 and 9.2, the knowledge base in which problems are developed should be held as constant as possible so as to allow maximum variation in problem-solving processes. This includes familiarity with the vocabulary, its general use in the designated context, its specific use relative to the problems, and its use within the structure in which the problems are designed.

Problems should be developed within a context that is known and readily interpreted by the child. That is why we focused on barnyards and workers in the previous illustrations. Other contexts include city streets, classroom settings, and toy stores.

We have found that problem-solving activities can be developed within any knowledge base (Cawley 1984b). In fact, there is considerable potential in utilizing the knowledge base contained in basal reading programs. Once the child has read the story, mastered the vocabulary, and demonstrated comprehension of the content, one can build sets of problem-solving experiences from this base. These problem-solving activities represent an additional form of comprehension and serve to reinforce the reading.

The procedure illustrated in Figures 9.1 and 9.2 could be adapted to have the youngsters create the story mats and the object cards. The teacher could follow the same script by simply substituting the information in the materials developed by the students.

Prerequisite Skills

Developmental or remedial programs, such as those considered by Blankenship and Lovitt (1976), stress reading as a prerequisite to problem solving. Figure 9.3 lists seven steps to follow in problem solving. Reading is only a prerequisite to problem solving when the problems are written.

As shown in Figures 9.1 and 9.2, problem solving can be conducted in a multimodal format where reading can be included or excluded according to

To solve a word problem, follow these seven steps.

1. Read the problem carefully.
2. Decide what question the problem is asking.
3. Decide what facts in the problem can be used to answer the question.
4. Decide what operation to use.
5. Write a number sentence that will help you solve the problem.
6. Reform the number sentence and write the answer to the problem.
7. Examine your answer, then check it.

FIGURE 9.3   Steps in solving word problems

the present level of functioning of the child. When the primary concern is mathematical problem solving, it seems that the mildly handicapped child should be given every opportunity to function at his or her maximum level of functioning in problem solving. Accordingly, if the child is a poor reader or reads significantly below the level required in the problem, the reading requirements may interfere with the child's ability to get to the problem in order to solve it. Ballew and Cunningham (1983) demonstrate that this is not a problem unique to the mildly handicapped. They found that only 12 percent of 217 sixth-grade children could read and set up problems at a level higher than they could compute, whereas 60 percent could compute correctly at a level higher than they could read and set up the problems.

A second prerequisite is computational ability. Again, we argue that when problem solving is the focus, computational requirements should not be so complex as to preclude the derivation of the correct solution due to an error in computation. At the very least, the youngster ought to be given two scores, one for correct problem solving and one for correct computation. Calculators should be utilized in problem solving.

Processes

We have committed ourselves to the development of comprehensive word-problem programs. Therefore, we feel one should designate the specific sentence structure or logical characteristics of each problem. That is, we propose to stipulate that a problem either requires categorization or does not, that it contains extraneous information or does not, that it is written in simple or complex sentences or is not. Whatever the characteristics of the problem, we suggest that they be stipulated before the problems are prepared. The remainder of this chapter will present word problems along with a brief description of the characteristics of each set of problems. Figure 9.4 lists some of the factors that might be included in word problems of different types.

In developing the following illustrations, we will use one topic—workers and their resources. The problems will focus on farming and farmers, although any topic could be used. To make adaptations to another topic, one need only substitute the appropriate terms. The term *car* could be substituted for tractor. Terms such as *bears* and *tigers* could be substituted for *apples* and *pears*.

SOME CONDITIONS OF WORD PROBLEMS

Classification

Vocabulary level and classification are two factors to consider when developing word problems. Figure 9.5 contains illustrations of these two factors and how they are used in combination with one another.

1. Set complexity
2. Neutral questions vs. direct questions
3. Direct vs. indirect
4. Continguity/noncontiguity of information:
    4.1 Single sentence
    4.2 Multiple sentences
5. Extraneous
    5.1 Verb
    5.2 Object
    5.3 Subject
    5.4 Phrase
    5.5 Clause
6. Cloze
    6.1 Multiple choice
    6.2 Open-ended
7. Conjunction, negation, class inclusion
    7.1 Conjunction/disjunction
    7.2 Negation
    7.3 Class inclusion
8. Syntax
    8.1 Simple sentence
    8.2 Structures of modification
        8.2.1 Prepositional phrase as noun, verb, or sentence modifier
        8.2.2 Complex sentence where relative clause modifies subject
        8.2.3 Complex sentence where relative clause modifies object
    8.3 Structure of predication
        8.3.1 Passive voice in simple sentence
        8.3.2 Passive voice in complex sentence
            8.3.2.1 When relative clause contains passive voice
    8.4 Structures of complementation
        8.4.1 Direct object/indirect object sequence
        8.4.2 Direct object/objective complement sequence
    8.5 Structures of coordination
        8.5.1 Sentences with coordination of phrases
        8.5.2 Compound sentences
    8.6 Combination of structures
9. Indefinite/definite quantifiers
10. Missing elements
11. Multiple step and combinations
12. Order of question
13. Position of quantity
14. Pictures in multiple choice to information
15. Information to multiple-choice pictures
16. Semantic: multiple meaning of word with question to one meaning
17. Meaningful vs. nonmeaningful
18. Nonoperation questions

FIGURE 9.4 Factors used in developing word problem experiences

Classification is a term that refers to the organization and superordination of the components of the information set. Simple subject (SS) means that the subject of the sentence is the same in each information statement and in the question (e.g., worker). Complex subject (CS) means the subject of the sentences changes in each information statement (e.g., farmer, driver) and that they lose their original identity in the question. They are linked together by their relationship in a specific concept (e.g., farmers and drivers are referred

	Vocabulary: Easier Set Complexity		Vocabulary: Harder Set Complexity
SS/SO	A worker had 3 trucks. Another worker had 2 trucks. How many trucks did the workers have in all?	SS/SO	A biologist illuminated 3 micro- scopes. Another biologist illuminated 2 microscopes. How many microscopes did the biologists illuminate in all?
SS/CO	A work had 3 trucks. Another worker had 2 tractors. How many vehicles did the workers have in all?	SS/CO	A biologist illuminated 3 micro- scopes. Another biologist illuminated 2 spectroscopes. How many pieces of equipment did the biologists illuminate in all?
SC/SO	A driver had 3 trucks. A farmer had 2 trucks. How many trucks did the workers have in all?	CS/SO	A biologist illuminated 3 micro- scopes. A chemist illuminated 2 micro- scopes. How many microscopes did the scientists illuminate in all?
CS/CO	A driver had 3 trucks. A farmer had 2 tractors. How many vehicles did the workers have in all?	CS/CO	A biologist illuminated 3 micro- scopes. A chemist illuminated 2 spec- troscopes. How many pieces of equipment did the scientists illuminate in all?

FIGURE 9.5   Vocabulary and set complexity in word problems

to as *workers*). The same is true for simple objects (SO) and for complex objects (CO). In effect, we can systematically manipulate the classification of subjects and objects within problems of the same type. Word classes (e.g., verbs, subjects, adjectives) have been shown to have different influences on problem solving with the mildly retarded (Penner 1972). The classification of verbs creates the most difficulty, as illustrated in an item where *bought* and *purchased* describe the actions in the information sets, but *obtained* is the word used in the question. If one farmer bought a truck and another farmer purchased two trucks, how many trucks were obtained by the farmers?

It is important to determine the relative influence of the effects of vocabulary and classification. That is, one might conclude that the learner has difficulty organizing the concepts (e.g., chemist and biologist = scientist) when in fact the use of an easier vocabulary might show that this is not the case. The difficulty could be one of vocabulary knowledge, not one of classification.

Extraneous Information

Extraneous information is another important influence in word problems. In fact, extraneous information has been the most studied of all word-problem components among the mildly handicapped, for example, by Cruickshank (1948) and by Goodstein (1974). Extraneous information is information that is irrelevant to the problem. It is information that if used will produce an incorrect response. Confusion with the topic of extraneous information is rampant. For example, in *Key Math Diagnostic Arithmetic Test* (Connolly, Nachtman and Prichett 1971), a problem is presented as follows:

Examiner says, "Six children are playing in a park that has 8 swings. If only 2 children are swinging, how many swings are empty?"

The behavioral objective for J-6 states: "Given a subtraction problem containing extraneous information, [the learner] solves for differences."

Four children were tested and the following responses were given:

Child 1: "Six"
Child 2: "Six"
Child 3: "Six"
Child 4: "Five"

When each child was asked to explain how the answer was obtained, the following responses were provided:

Child 1: "I subtracted 2 from 8 and got 6."
Child 2: "I added the 4 here (merry-go-round) to the 2 here (swings) and got 6."
Child 3: "I said 6 because I heard you say 6."
Child 4: "I subtracted 2 from 8 and got 5."

Notice that Child 1 and Child 4 solved the problem as intended. Each was a good problem solver; each subtracted and did not pay attention to extraneous information. Child 1 was given credit for a correct response, and Child 4 was scored as having given an incorrect response. Yet, neither Child 2 nor Child 3 actually "solved" the problem. Each used extraneous information in a form stated by the examiner or in the form of a visual analysis of the children in the park. Each of these was given credit for a correct response, even though neither solved the problem and both were foiled by the extraneous information. A better picture would have shown 5 children on the merry-go-round and 2 children on the swings. A response of *seven* would have alerted the examiner that the child was attending to the visual display and not to the oral conditions. Further, the examiner should not have stated *six*. The examiner should have stated that *some children* were playing.

Key Math problem J-9 is presented as follows:

Examiner says, "Mr. Drake hires 3 boys to work in his store. He pays each boy $1 per hour. John worked 6 hours, Bill worked 11 hours, and Tim worked 7 hours. How much more money did Bill earn than John?"

The behavioral objective for this item states, "Given a problem requiring two computations, [the learner] solves by subtraction and multiplication."

The behavioral objective makes no reference to extraneous information. Yet, *Tim worked 7 hours* certainly qualifies as extraneous information. Should Tim be used in place of either John or Bill, the answer would be unacceptable. Further, the item does not require multiplication because *one* is the identity element of multiplication, and each factor would retain its own identity. Paying the boys two dollars per hour would have required multiplication and made the item much better for diagnostic purposes.

Types of extraneous information. Two types of extraneous information can be used in problems. One type manipulates a qualitative distractor and the other type manipulates a quantitative distractor. A qualitative distractor is a part of the information set, such as an object, a phrase, an adjective, a verb, or any other information statement that is irrelevant. A quantitative distractor is a quantitative property, such as a cardinal number that is mentioned in the information but is not integrated into the computation when performing the operation. Qualitative distractors can be integrated into the problem in the form of words, objects, or pictures. Quantitative distractors require the use of pictures or objects. Let us see how this works.

No distractor	*Distractor*
One farmer has 3 trucks.	A farmer has 3 trucks.
Another farmer has 2 trucks.	Another farmer has 2 trucks.
Another farmer has 4 trucks.	Another farmer has 4 tractors.
How many trucks do the farmers have in all?	How many trucks do the farmers have in all?
A farmer has 3 tractors.	A farmer has 3 tractors.
A driver has 2 trucks.	A driver has 2 trucks.
A gardener has 4 trucks.	A gardener has 4 rakes.
How many vehicles do the workers have in all?	How many vehicles do the workers have in all?

The first example of a no-distractor item mentions only farmers and trucks and asks about all of them. The information in the first distractor problem mentions farmers, trucks, and tractors, but asks only about trucks. Tractors represent extraneous information.

The second example of a no-distractor problem mentions the set of workers and a set of vehicles and asks about both. The information set of the second distractor problem mentions vehicles, workers, and rakes, but asks only about workers and vehicles. Rakes constitute extraneous information.

The first pair of items did not focus on classification, whereas classification was integrated into the second set of problems. Thus, we suggest that classification type activities precede all others. Problems containing quantitative distractors must be accompanied by pictures or objects. Figure 9.6 shows pictures of farmers and a gardener.

The information given to the learner in Figure 9.6 is:

A farmer has 3 tractors.
A gardener has 4 riding mowers.
How many pieces of equipment do the workers have?

There is no mention of *twoness.* In this instance, twoness is a quantitative distractor and is extraneous. The learner must attend only to the relevant quantitative information and exclude twoness even though it is presented as part of the set of workers and equipment.

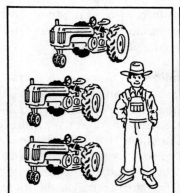  

FIGURE 9.6

FIGURE 9.7

In Figure 9.7, we examine a task that contains both a quantitative distractor and a qualitative distractor. The information presented to the learner in Figure 9.7 is:

A farmer has 3 tractors.
Another farmer has 2 trucks.
How many vehicles do these farmers have together?

There is no mention of the farmer with two tractors, thereby making the information in that picture a quantitative distractor. There is no mention of the gardener or the rakes, thereby making each of these qualitative distractors.

Many feel that the use of pictures or objects in the problem-solving setting helps to increase attention to relevant information. To a degree this is true (Schenck 1973). However, this is not an all-inclusive generalization. Some children will continue to have difficulties even when pictures or objects are utilized. This is particularly so in extraneous-information problems where the teacher mentions all the information. What happens is that the child does not use the pictures. He or she simply listens to the information and computes as though the pictures were not present.

One way to attack this problem is to replace the number statement with a word such as *some, few, bunch,* or *group* and allow this term to serve as an indefinite quantifier (Schenck 1973). Pictures such as the ones in Figure 9.7 can be presented with an information statement such as:

Some farmers have some tractors.
A farmer has some trucks.
A gardener has some rakes.
How many vehicles do these workers have in all?

Problem Solving: Word Problems

One can see that the term *some* serves only to direct the attention of the child to the fact that the items mentioned in the object position of the sentence have a quantitative property. The child needs to search the pictures to locate the needed information and to determine the "how manyness" associated with each. This form of forced search focuses attention on the relevant information and helps the child to reject extraneous information.

Question Types

Instructional activities with word problems continuously emphasize the role of the question. In fact, children are often told to look at the question as though it is more important than the set of information to which the question is related. Most questions contain cue words, and the child is often instructed to do what these cue words say. Such is shown in the following:

Cued questions

A farmer had 3 tractors.

The farmer bought 3 more tractors.

How many tractors does the farmer have in all?

A farmer had 3 tractors.

The farmer sold 2 tractors.

How many tractors does the farmer have left?

The terms *in all* and *left* signal the child to perform an operation, and the child does so without analyzing or interpreting the information set. The question is the driving force. As we shall see momentarily, dependence on cue words sidetracks the child from the information set and leads to rote responses. An alternative to the cued question is the neutral question. Neutrally stated questions do not signal or cue any operation. Their main purpose is to indicate that it is time to examine the information and develop a response.

Neutral questions

A farmer had 3 tractors.

The farmer bought 2 more tractors.

How many tractors does the farmer have now?

A farmer had 3 tractors.

The farmer sold 2 tractors.

How many tractors does the farmer have now?

As can be seen, the question is of no help to the child. He or she must read and analyze the information, determine that the verbs *bought* and *sold* signal different operations, and then associate the proper action with each verb. Thus, the importance of language, its vocabulary, and its structure begin to take on more meaning in problem solving. The child is required to become more cognitively active and to exercise more self-control over the metacognitive processes of searching, planning, and executing.

Direct and Indirect Problems

Direct problems are those in which the language meanings of the information set are consistent with the operation one would commonly use to obtain a response. Indirect problems are those problems in which the language meanings are inconsistent or in contradiction with common usage in word problems. There are different interpretations as to the preferred ways to state indirect problems. For purposes of this chapter, we will use the term *indirect* to mean that the operation to be used is the opposite of that which is commonly associated with the terms. For example, terms related to *addition* will be used in problems that require *subtraction*.

Direct	*Indirect*
Addition	*Addition*
A farmer has 3 tractors. The farmer added 2 more to her farm. How many tractors does the farmer have now?	A farmer added 3 tractors to her farm. The farmer now has 5 tractors. How many tractors did the farmer start with?
Subtraction	*Subtraction*
A farmer had 5 tractors. The farmer sold 2 tractors. How many tractors does the farmer have now?	A farmer had 3 tractors left after 2 tractors were sold. How many tractors did the farmer start with?
Multiplication	*Multiplication*
A farmer plowed 6 acres with each of her 3 tractors. How many acres did the farmer plow?	A farmer plowed 18 acres. This is 3 times as many as one tractor can plow. How many acres can one tractor plow?
Division	*Division*
A farmer had 18 rows to plow. She plowed 6 rows of corn with each tractor. How many tractors did the farmer use?	A farmer divided the plowing so that each tractor plowed 6 rows of corn. The farmer had 3 tractors. How many rows did the farmer plow?

As can be seen in the indirect items, terms such as *added, times, divide,* and *left* are utilized in a manner quite contradictory to that found in problems that accentuate the use of cue words. We believe that indirect problems need to become integral components of any problem-solving program for mildly handicapped children. To minimize the extent to which they will have to "unlearn" inappropriate behaviors, it is our suggestion to delete problems that stress the search for cue words in questions. Better to begin all problem solving with neutrally stated questions and have the child become accustomed to searching, analysis, and interpretation from the very beginning.

Cloze Procedure

Cloze is a procedure that requires the learner to insert words in spaces where terms or words have been deleted. The cloze method is most commonly used in reading and language activities. The use of cloze procedure is consistent with our definition of word problems, because cloze activities require the analysis and interpretation of information. Numerous activities can be developed by varying the cloze approach across qualitative and quantitative conditions.

Cloze procedure may be used in either multiple-choice or open-ended formats. When used in multiple-choice formats, the teacher can exercise complete control over the words, the number of choices, and the quantitative characteristics. A smaller number of choices (e.g., two versus four) increases the probability that a correct response will occur. Thus, for a child who is not having a great deal of success, the two-choice option is a good one.

Open-ended formats enable the teacher to determine more about the vocabulary of the children. The teacher can make a reasonable determination of the range of vocabulary available within the group and use this to create other problems. One approach might be to use the same words in situations where their meanings differ. Such an example was illustrated with the indirect problems.

Multiple-choice formats can be used to introduce new terms and to give children an opportunity to use them in varying contexts.

Multiple-choice formats. Mark the word (verb) that makes each set of information true.

A farmer had 3 tractors.　　　　　　　A farmer had 3 tractors.

The farmer $\begin{bmatrix} \text{sold} \\ \text{bought} \\ \text{painted} \end{bmatrix}$ 2 tractors.　　The farmer $\begin{bmatrix} \text{sold} \\ \text{bought} \\ \text{painted} \end{bmatrix}$ 2 tractors.

The farmer has 5 tractors.　　　　　The farmer has 1 tractor.

Mark the word (object noun) that makes each set of information true.

A farmer has 3 tractors.　　　　　　A farmer has 3 tractors.

The farmer bought 2 $\begin{bmatrix} \text{trucks} \\ \text{tractors} \\ \text{rakes} \end{bmatrix}$.　　The farmer sold 2 $\begin{bmatrix} \text{trucks} \\ \text{tractors} \\ \text{rakes} \end{bmatrix}$.

The farmer has 5 tractors.　　　　　The farmer has 1 tractor.

Open-ended cloze. Write in a word that makes each set of information true.

A farmer has 3 tractors.　　　　　　A farmer has 3 tractors.
The farmer (　　　　) 2 tractors.　　The farmer (　　　　) 2 tractors.
The farmer has 5 tractors.　　　　　A farmer has 1 tractor.

Write in a word that makes each set of information true.

A farmer has 3 vehicles.　　　　　　A farmer has 3 vehicles.
The farmer bought 2 (　　　　).　　The farmer sold 2 (　　　　).
The farmer has 5 vehicles.　　　　　The farmer has 1 tractor.

The cloze task can be extended to include more than one deletion in each information statement. When this is the case, children are constructing problems that can then be given to others.
To illustrate:

A _____ has 3 ____ .　　A _____ has 3 _____ .
The _____ ____ 2 ____ .　　The _____ ____ 2 _____ .
The _____ has 5 ____ .　　The _____ has 1 _____ .

Missing-Element Problems

Missing-element problems require the individual to search a body of information, relate this information to a statement or a question, and then determine the adequacy or inadequacy of the information for the problem. Missing-element problems can be developed to focus on different language or cognitive needs. That is, they may vary in vocabulary, classification, or any

71. See the pictures of the animal cages.

How many more bears than lions and tigers?

Point to the missing information:

_____ A. The number of tigers in the 3 cages.

_____ B. The number of bears in the 3 cages.

_____ C. The number of lions in the 3 cages.

_____ D. The number of monkeys in the 3 cages.

72. A group of 24 children went to a zoo.
Another group of 16 children went to the same zoo.
How many boys went to the zoo?

Point to the missing information:

_____ A. The number of children who went to the zoo.

_____ B. The name of the zoo.

_____ C. The number of girls who went to the zoo.

_____ D. The cost of each ticket.

FIGURE 9.8   Missing-element problems

other trait. Missing-element activities can be conducted with pictures, objects, or with specific roles that children and teachers play (see Figure 9.8).

Missing-element problems may include problems in which qualitative information is missing or problems in which quantitative information is missing. Qualitative problems would be those in which there is incomplete information concerning the subject, object, verb, a phrase, or any other comparable item, as shown below:

There are 3 tractors.
There are 4 trucks.
There are 2 rakes.
How many vehicles do the farmers have in all?

As can be determined, there is no indication that the items belong to a farmer. They could be the property of anyone.
Examine this item:

John has 2 toy tractors.
Kate has 3 toy trucks.
Her friend has 1 toy car.
Jane has 2 toy tractors.
How many toys do the girls have in all?

The term *her friend* does not indicate if it is a boy or a girl. The youngster must ask for clarification.

Quantitative missing-element problems are those types in which there is a lack of needed quantitative information, as shown in the following:

A farmer has some tractors.
Another farmer has 3 trucks.
How many more tractors than trucks?

A farmer has some tractors.
Another farmer has 3 trucks.
Are there more trucks or more tractors?

Grammatical Considerations

The previous discussion focused on problem types where the emphasis was on vocabulary and ways in which vocabulary could be integrated into word problems to vary the information-processing requirements. This section will focus on the structure of our language and provide illustrations to show how problems can be written in different syntactical arrangements.

Trenholme, Larsen, and Parker (1978) studied the effects of different types of sentence structure on the solving of word problems.

Breault (1983) studied the performance of learning-disabled children on problems that varied in both syntax and in the use of extraneous information. In the Breault study, the effects of the extraneous information tended to be more pervasive than the effects of sentence complexity.

The way we use words in combination with one another and the way we sequence word combinations create various meanings. For example, in the sentence "John gave Katie 4 apples," we have John as a giver and Katie as the receiver. In the sentence "Katie gave John 4 apples," we have Katie as the giver and John as the receiver. If we change the words and make the sentence "John was given 4 apples by Katie," we now have John the subject as the receiver and Katie the object as the giver. In effect, by bringing certain language conditions into our problem solving, we create divergent types of problems and provide children with an opportunity to interact with different language structures.

In this section, we will continue with our theme: Workers and their resources. We will keep the vocabulary as constant as possible and build illustrations of problem activities by changing or enlarging upon structure, as shown below:

Subject	Verb	Direct object
The farmer	ate	3 apples.
Another farmer	ate	6 apples.

How many apples did the farmer eat in all?
This first illustration shows a simple sentence in which a subtraction-type verb (loss of) is used in an addition problem.

Subject	Verb	Indirect object	Direct object
The farmer	sent	her daughter	2 apples.
The farmer	sent	her son	3 apples.
The farmer	sent	his daughter	4 apples.

How many apples did the farmers send to their daughters?

This second illustration maintains a relatively simple sentence structure, but provides the potential for some relatively complex information-processing conditions (e.g., classification, extraneous information).

Subject	Verb	Object	+	Subject	Verb	Object
Helen	picked	4 apples	and	Marie	picked	2 apples.
Jose	picked	4 oranges	and	Tim	picked	2 oranges.
Albert	picked	3 carrots	and	Angie	picked	3 pears.

How many pieces of fruit were picked by the boys?

These examples of compound sentences combine two simple sentences; they require classification and the exclusion of two sets of extraneous information (i.e., Albert, who picked carrots; and Helen, Marie, and Angie, who are girls).

Subject	Verb	Indirect object	Direct object	Modifier
The farmer	sold 3	of the apples	to the boy	behind the counter.
The farmer	sold 2	of the pears	to the boy	behind the counter.
The farmer	sold 3	of the apples	to the girl	behind the counter.
The farmer	sold 2	of the apples	to the boy	in front of the counter.

These sentences include a prepositional phrase as a modifier and use this and all other components except the subject as potential sources of extraneous information. There is considerable information in these four sentences. Should they be given orally, the teacher would have to be concerned with the demands of short-term memory and with the need for the learner to pay attention. In such a circumstance, it would seem that one would interact extensively with the learners and ask numerous questions, rather than asking one question and then giving another set of information. One might ask questions such as the following:

1. "How many apples were sold to the boy?" This question creates a single-digit addition task. The primary sets of extraneous information are *girl* and *pears*.

2. "How many apples were sold to the boy in front of the counter?" This is a simple location item in which there is no operation other than locating the number in front of the counter. This is tricky information in that it does not tell us if the boy was alone in front of the counter or whether the farmer and the boy were in front of the counter at the same time.

True, False, or Can't Tell

The preceding question contains an element of ambiguity. This was done to allow us to move into another problem-solving condition that we refer to as *true, false,* or *can't tell.* This type of activity involves extensive information analysis, considerable attention, and a capacity to tolerate ambiguity. Using the same information previously presented about the farmer, the boy and the girl, the apples and pears, and the positions in front of or behind the counter, the learner is asked to respond to each of the following with a True, a False, or a Can't Tell.

1. The boy is behind the counter.
2. There are 3 apples behind the counter.
3. The farmer sold only 5 fruits.
4. The farmer sold only 8 fruits.
5. All the fruit are behind the counter.
6. The farmer sold 10 fruits.
7. There are four fruit counters.
8. The boy bought more fruit than the girl.
9. The farmer likes apples better than pears.
10. The farmer sold all her fruit.
11. The farmer has 3 fruits left for herself.

Each youngster could mark his or her response to each item. The teacher could collect data in the form of a poll and determine the number of varying responses to each item. A discussion could take place in which different children defended or described their responses. The teacher could indicate a need to clarify the *can't tell* responses. Objects could be given to individual children, and they could be instructed to place the objects in positions such that the *can't tells* are changed to either true or false. Again, a poll needs to be taken because some children will make things true that other children will make false. It is important to realize that either could be correct. If objects are not available, the children could draw pictures.

This next example uses the true, false, or can't tell conditions in a complex sentence where the relative clause modifies the object.

The accidents

The 3 tractors hit the tree that the farmers were cutting.

Another 2 tractors hit the pole that the farmers were fixing.

The 4 trucks hit the tree that the farmers were cutting.

Mark the following true, false, or can't tell based on the information given to you.

1. The farmers were cutting a tree that was hit by some vehicles.
2. A pole was being fixed by some men.
3. At least 5 cars hit things the farmers were fixing.
4. The farmers were cutting the cars.
5. The tractors were hitting the pole for the farmers.
6. More tractors hit the tree than hit the pole.
7. At least 5 vehicles hit the tree that the farmers were cutting.
8. There were at least 9 vehicles in the accident.
9. More vehicles hit the pole than hit the tree.
10. No damage was done to the van.
11. There were 3 more tractors than trucks in the accident.

As can readily be determined, numerous issues need clarification. In Item 2, for example, there is no evidence to determine that the farmers were men. Item 5 requires classification of the tractors and trucks into a set of vehicles. Item 10 indicates that no damage was done to the van. What van? Again, it is possible for children to play roles, manipulate objects, draw pictures, or to negotiate responses in discourse.

The final format to be described in this chapter will be that of complex sentences in which the information units and the questions will be presented in distinct units and those in which the information units and questions are linked together in one statement. The first set presents two examples in which the question is distinct from the information.

Mr. Henning had 3 tractors before he bought 2 more.

Mr. Clark had 4 trucks before one was wrecked.

Mr. Johnson had 3 tractors after he bought 1 new tractor.

1. How many tractors does Mr. Henning have in all?
2. How many trucks did Mr. Clark have after the wreck?
3. How many tractors did Mr. Johnson have after he bought a new one?

Red apples, because the price was raised 2 cents each, now cost 9 cents.

Yellow apples, although they will go up 1 cent each next week, now cost 7 cents each.

Green apples now cost 8 cents each because the price dropped 1 cent each.

1. What was the price of the red apples before the price increase?
2. What will yellow apples cost after the price is increased?
3. What was the price of green apples before the price was dropped?

Each of the illustrations makes use of clauses that control time-event (i.e., before/after) conditions that the learner must interpret in order to seek the appropriate operation. These types of activities can be conducted with groups, in which children can assume different roles and act out the events. Acting the role of a farmer purchasing tractors can be considerably different from acting the role of an apple going down in price. Yet each provides an opportunity for active involvement, attention to detail, and the sequencing of actions to produce an acceptable response.

A somewhat different format presents the item in a manner in which the information and question are contained within a single information set, as illustrated by:

If a farmer bought 2 tractors to go with the 3 she had, how many would she have altogether?

If a farmer bought 2 tractors and another farmer bought 2 tractors and a truck, how many tractors would they have altogether?

How many tractors would the farmer have after buying 2 to go with the 3 she already owned?

The following use the same format to illustrate three basic subtraction questions. One question is "What must be added to a number to make it as large as another number?" The problem might read:

If a farmer wants 5 tractors, but has only 3, how many would she have to buy?

A second question is, "What is left of a quantity after part of it has been taken away?" A problem might read:

If a farmer had 5 tractors and sold 2 of them, how many tractors would the farmer then have?

A third question is, "How much larger is one number than another?" A problem might read:

> If one farmer has 5 tractors and another farmer has 2 tractors, how many more tractors does the first farmer have than the second?

SUMMARY

This chapter has only presented activities that were representative of word-problem types and some conditions that could be used to construct word-problem experiences for mildly handicapped children. The focus was on workers and their resources, although any topic of interest could be used. Instructors and learners can work together to create materials and to act out various roles and requirements. Word problems are viewed as active learning experiences for children. We caution against the traditional and passive classroom activity that focuses on the workbook page and its 8–10 items that seek only the correct answer through cue-word dependence. We advocate alternative modalities in the word-problem lesson. Mildly handicapped children need to be active, not passive learners; they need to be able to deal with problems, not repetitive exercises; they need to be able to access the problem, not be bogged down by excessive reading and computational demands that detract from problem solving; they need to have frequent, if not daily, experiences with word problems; and they need to be *taught* during the problem-solving stage, *guided* during the exercise stage, and *motivated* to perform quickly and accurately during the example stage. Our goal is to have mildly handicapped children solve all problems at the exercise stage.

PROBLEM SOLVING: DIVERGENT CONSIDERATIONS

The preceding chapter focused on word problems. In the majority of illustrations, the conditions of word problems were manipulated to meet a variety of learner needs and to provide a divergent view of word problems. The emphasis was on correct responses. The outcomes were correctness, acceptability, and reliability. This chapter will present a set of different strategies for conducting problem solving. The emphasis will be on creative and constructive perspectives of problem solving and will highlight original thought, personal evaluation, and the long-term interaction with information and data. This chapter also includes a section on appraisal and a section on instructional procedures.

Figure 10.1 displays an overview of the dimensions of problem solving as developed in this text. The items to the left stress solutions, and the goal is for the child to produce correct, reliable, and acceptable responses. The items to the right stress the creative aspects of problem solving and seek originality, flexibility, and long-term personal involvement with information and data that can be used for decision-making purposes or for the development of arguments. The term *argument* is used to mean the following: to justify one's position, to take a point of view that can be substantiated from a data base, or to challenge the conclusions of persons or systems.

Three formats of problems emerge. First, we have the traditional word-problem format consisting of three or four lines of information that are followed by a question. Second, we have information that is presented in the form of a display that shows information in the format of a graph or figure. Third, we have information that is presented in story form. Figures 10.2 and 10.3 show illustrations of display and story formats. Variation among the three sets of formats is essential in problem-solving programs for the mildly handicapped, for they involve children in information and data that are more similar to what is experienced in real life; for example, newspaper ads that are forms of displays, newspaper stories that are comparable to story formats.

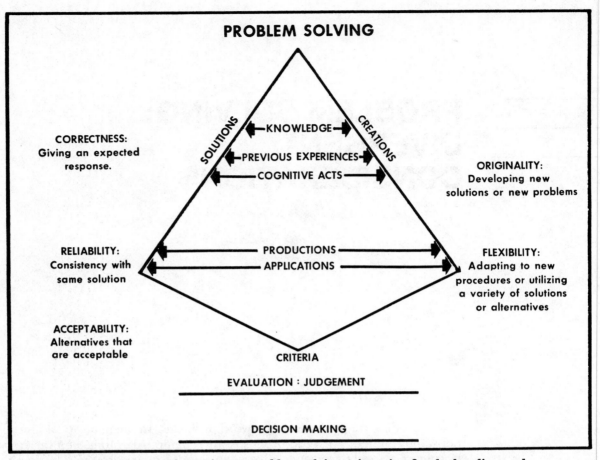

FIGURE 10.1 Alternatives to problem-solving orientation for the handicapped

PROBLEM-SOLVING FORMATS

Constructing Word Problems

Three forms of problem activities fall under the heading of construction. The first type provides a set of information and requests the learner to make a set of questions relevant to the information. To illustrate:

A farmer has 3 tractors.
Another farmer has 2 tractors.
Another farmer has 4 trucks.

Look at the sentences.
List all the questions you can think of that go with the sentences.

A second type of construction begins by providing the youngster with sets of questions. The youngster is required to prepare sets of information that will make it possible to answer the questions.

_____
_____
_____
_____

How many more tractors than trucks?
How much farther did the farmer drive than the gardener?

PIECE WORK

Many people who work are paid a salary for each week of work. Other people are paid by the hour. Still other people are paid for the work they do such as in assembling calculator parts. Each person is paid by the piece — that is, for each calculator that is assembled. The rates are as follows:

Job No. 1. electronic assembly..9¢ per unit
Job No. 2. putting electronic assembly in case..........................7¢ per unit
Job No. 3. packing calculator in shipping box...........................4¢ per unit

	Amount Earned	
	Monday	Wednesday
Harold		
Job No. 1	22.23	14.85
Job No. 2	13.65	16.52
Job No. 3	9.32	11.00
Octave		
Job No. 1	36.09	21.33
Job No. 2	3.57	11.76
Job No. 3	7.32	5.80
Beth Mae		
Job No. 1	16.38	32.76
Job No. 2	--	11.38
Job No. 3	29.76	2.60

1. How many electronic assembly units did Harold do on Monday?
2. How many electronic assembly units did Harold do on Wednesday?
3. How many electronic assembly units did Octave do on Monday?
4. How many electronic assembly units did Octave do on Wednesday?
5. Beth Mae did not do one type of work on Monday. What type was it that she did not do?
6. How many calculators did Beth Mae put in a shipping box on Wednesday?
7. How many calculators did Beth Mae put in a shipping box on Monday?
8. How many electronic assembly units did Octave put in a case on Wednesday?
9. How many electronic assembly units did Beth Mae put in a case on Wednesday?
10. How many calculators did Harold put in shipping boxes on Wednesday?
11. How many electronic units were put in cases by Octave on Monday?

FIGURE 10.2 Display format for problem solving

A third type of construction activity provides the child with a set of words and a set of numbers. The child is required to use the words and the numbers to create problems.

Words	*Numbers*
farmer, gardener	9,7,4
lettuce, corn	
turkey, pork	

See the words and the numbers.
Use the words and the numbers to make as many different
  problems as you can. You may use other words. Write
  complete sentences.

The different problems that are generated by the children can be given to other children to judge, to complete, and to modify. They can be grouped by the different operations or by their characteristics. Cohen and Stover (1981) gave children problems that contained extraneous information and asked the children to rewrite the problems to make them easier. The children removed the extraneous information. The teacher could determine the extent to which

An exciting airplane. Yes indeed, it was an exciting airplane. That's what everyone said. "That seven-forty-seven is an exciting airplane." The one we were on carried 342 passengers. When all those people went to get their luggage, there was a big mix-up. The mix-up was really due to the bad weather that forced five 747's to land at the same time. Ours, the United with 342 passengers, was the second to land. Pan American had landed first with 339 passengers. TWA was third with 301. Northwest landed fourth with 317 passengers on board, and American Airlines came in fifth with 314. Everybody decided to let the airline staff organize the groups; otherwise, there would be a mess.

1. Which airline carried the fewest number of passengers?
2. An airline with 301 passengers landed third. Which airline was this?
3. How many passengers were on the first two planes to land?
4. How many passengers were there in all on the last three planes to land?
5. What was the total number of passengers on all five airplanes?
6. Northwest and Pan American shared a baggage terminal. How many people would be going there for bags if everyone had at least one bag?
7. One passenger thought the planes should land in sequence, with the one carrying the largest number of passengers landing first. What would have been the order had this been done? List the airlines by name in the order suggested.

FIGURE 10.3   Story format for problem solving

children developed problems with extraneous information or the extent to which they prepared problems with neutral questions. Children could be eased into these tasks by focusing on one characteristic at a time. That is, the instructor could set the limits by requesting that the children prepare problems that include specific characteristics (e.g., extraneous information).

Story Problems

We have shown that word problems can be written to nearly any set of specifications. As such, word problems can be incorporated in activities in language, reading, cognition, and nearly any other need system of the mildly handicapped. The primary characteristic of the traditional word problem is that it contains 3 or 4 lines of information that are followed by a single question. As such, the reading is abrupt and lacks continuity across a theme.

A variation on the word problem is the story problem (Cawley et al. 1974, 1976). Story problems range in length from single paragraphs to full-length stories. They may be taken from available reading material (Cawley 1984b), or they may be specifically constructed to provide applications in mathematics, as in Figure 10.3. Story problems bring together a number of pieces of information, highlight a theme, and represent an activity that is common to school programs. The incorporation of story problems into the school program provides an excellent opportunity for a youngster to read material much below his or her age level. This is done under the guise of problem solving and reduces the uncomfortable feeling many youngsters have when confronted with material they know is associated with lower grade levels. The early experiences of the child are extensively quantitative. On a page-for-page basis, "The Three Little Pigs" is likely to contain as many quantitative references as any given source. Variations on the theme can be presented by asking the child to tell or write a story of two little pigs or of four little pigs. Comparisons can be made of the manner in which they use quantity to vary context, vocabulary, or sentence characteristics.

Fairy tales that begin with "Once upon a time" introduce children to time-and-event relationships by using the introductory statement to connect the present with the past. Story problems offer another adjunct to the program for the mildly handicapped in that they also serve as additional lessons on reading comprehension. Stories can be modified or written for any topic at any level desired by the instructor. Questions and other comprehension activities can be formulated from differing perspectives. There can be a balance between literal and interpretive questions. A story can be selected because it meets the needs of a learner in specific reading skills.

An example of the latter can be found in "Blue Flower" (McCracken and Walcutt 1963). The beginning levels of the Lippincott Basic Reading Program place great emphasis on the development of sound-symbol combinations in isolation, practice the sound-symbol combination with a set of words, and then present a story in which the sound-symbol combination is used in words in context. The story of "Blue Flower" is used to incorporate the long $\bar{e}$ sound of *ie* into words in context. The original story, found at the first-reader level, contains words and instances of *ie* correspondences. We have prepared the following modifications of "Blue Flower." In doing so, we have increased the number of *ie* correspondences through the use of quantifiers; for example, a pony was changed to nine ponies, and a flower was changed to daisies and pansies. We have also expanded the length and complexity of the story and developed two sets of questions, one quantitatively dependent and the other nonquantitatively dependent. The quantitative requirements are limited to counting and the addition or subtraction of single-digit combinations.

"Blue Flower"

There once was a little Indian girl named Blue Flower. Blue Flower was the niece of the chief of the tribe. His name was Sky Chief. Blue Flower lived with her mother and father. She had 3 brothers and 2 sisters. Her father had 7 ponies. Sky Chief had 9 ponies. Sky Chief and his wife had 4 children. Three of the children were girls. Sky Chief was fierce in battle. Sky Chief was also a gentle man. Sky Chief loved Blue Flower and her family and all the Indians in the tribe.

Nonquantitative questions

1. Blue Flower had an uncle. What was his name?
2. Why do you think Sky Chief was a good chief?
3. What kinds of animals did Sky Chief own?
4. How do you know that Sky Chief was an uncle of Blue Flower?

Quantitative questions

1. How many people in Blue Flower's family?
2. How many people in Sky Chief's family?
3. How many boys in Sky Chief's family?
4. How many of Blue Flower's cousins does the story tell about?

"Blue Flower"

Blue Flower is the niece of a fierce Indian chief. His name is Sky Chief. There were many stories about Sky Chief. The Indians told stories about his bravery on the field of battle. They told stories about his grief. They heard his cries when other braves were killed or wounded. Sky Chief was a gentle man. He loved the pansies and the daisies that grew near his tent. He loved his ponies and his puppies. Most of all, he loved his family. Sky Chief has 9 ponies. This is 2 more than his father has. Sky Chief has 4 puppies left after he gave 2 away to the children. Winter was coming and the tribe got ready to move. Blue Flower saw Sky Chief counting the flowers near his tent. He counted 6 daisies and 3 pansies. He picked 2 daisies to take on the trip. Sky Chief let out a shriek and the tribe knew it was time to move. Blue Flower and her 3 brothers and 2 sisters each climbed on a pony. The tribe headed south.

Nonquantitative questions	*Quantitative questions*
1. Name two things that make you think Sky Chief is a good chief.	1. How many ponies did Blue Flower's family own?
2. Why did the Indians go south?	2. How many ponies did Sky Chief's father own?
3. What word in the story tells you that the Indians knew it was time to go?	3. How many puppies did Sky Chief start with?

As can be noted from a review of the two stories, many more questions and problems can be created. In an actual teaching situation, we recommend that the teacher ask as many questions and give as many problems as possible. Our reasoning is that the children have become familiar with the story, its context, and its vocabulary orally or in written form and that it's wise to capitalize on this knowledge base for comprehension. Different levels or types of questions can be asked of different children in a group setting. Question variation is one means of adapting to individual needs.

Measure of comprehension should not be limited to questioning. One can prepare multiple-choice or true-false items. One can ask a child to rewrite the story or to paraphrase its meaning. Children can be asked to substitute key words or to interpret the meaning of certain words in other contexts. Children could be requested to act out certain roles.

Not all responses can come from memory. Thus, comprehension and problem solving should not be considered synonymously with retention or memory, although we recognize the general relationship between the two. Problem solving in particular requires that the child understand the question and search for the information needed to answer that question. In effect, the question becomes more of a driving force than the context.

It is important to help children with the development of different meta-cognitive and self-help skills in problem solving. When using the story format, the child must attend to the differences in types of questions that she or he has become accustomed to and those that are more novel. Children who are used to cognitive-memory type questions such as "What is the name of the chief?" or "What is the name of the girl in the story?" may require training and experience before dealing effectively with questions such as "How can you prove that Sky Chief was an uncle of Blue Flower?" or "Can you show and explain the part of the story that tells how many cousins Blue Flower had?"

Mildly handicapped children need the opportunity to experience different types of questions and problems and to be exposed to a variety of ways by which their comprehension can be demonstrated. The integration of reading, language, mathematics, and cognitive activities is one step in that direction (Cawley 1984b; Cawley et al. in press).

FIGURE 10.4 Indians as content source in problem solving

Displays

In addition to the use of the typical word problem and the story format, information can be presented in the form of a display. A display utilizes objects, pictures, graphs, or tables as the primary media.

The early use of the display (see Figures 9.1 and 9.2 as examples), provides opportunities for problem solving long before children read. The continued use of displays maintains the multimodal character of the information sets. Children can use three-dimensional objects to create displays, or they can draw figures and graphs to present data and information.

Figure 10.4 displays the quantitative information presented in the stories about Blue Flower. This display has the potential to interrelate other mathematics topics, such as measurement and geometry. For example, we could ask the child to look at the two Indian braves and name the people standing between them. This would bring the topological concept of *betweenness* into the lesson. Another topological concept, *order constancy,* could be addressed by asking the child to use the ordinal names (e.g., first) to explain and predict the constancy of the order in which the Indian children will ride south if they maintain their present arrangement. Should one have manipulative materials available, many different representations of the various concepts could be developed. There are some distinct advantageous to three-dimensional objects, particularly in the number of arrangements that can be made, the replicability of the same arrangements as needed, and the modification to more simple or complex arrangements as dictated by learner performance.

Problem Solving: Divergent Considerations

CREATIONS

The creations dimension of problem solving extends the range of problem solving and allows for much attention to individual interests and needs. The primary purpose of the creation dimension is to encourage originality in the development of alternatives to problem solving and problem-solving strategies. A second purpose is to encourage flexibility, the acceptance of new ideas, adaptation to new sets of procedures and tactics, and the active participation in the generation of alternative methods. The learner must learn to handle ambiguity, understand how different individuals are affected by it, and how one goes about resolving ambiguity. Although not all-important in finding solutions to word problems, originality, flexibility, and an ability to deal with ambiguity are extremely important when dealing with social problems.

Much of the emphasis in the solution dimension was in terms of attentional factors, organizational schemes, the reliability of procedures, closure, correctness, and appropriateness—all of which are taught directly or conducted under the guidance of the instructor. By contrast, the creation dimension stresses independence and the recognition that mildly handicapped children have creative and divergent capabilities. Mildly handicapped children are seen as being able to recognize problems, to develop plans for their solution, to collect and organize data, and to use these data in decision making or in argument. This does not imply that they can do all problems at all levels. There are developmental considerations that make it necessary to align the problem situation with the developmental status of the individual.

ACTIVITY

FIVE COMPONENTS OF THE CREATION DIMENSION AND RELEVANT ACTIVITIES

1. A problem is given. No solutions are provided. One or more are needed.
2. A problem is given. Alternative solutions need to be examined, choices made, or new solutions offered.
3. A problem is given. All solutions are ambiguous and lack closure.
4. A problem is to be created. Solutions need to be created.
5. Story production

1. A problem is given. No solutions are provided.
A. A teacher wants to buy some crayons. She has $8 to spend. One box has 12 crayons, 3 each of 4 different colors. One of these boxes costs $2. Another box has 3 each of 6 different colors. One of these boxes cost $3. The teacher wants 48 crayons. What should she buy with the $8?
How can we help the teacher decide what to buy?
Plan: Display the information as indicated in Figure 10.5. Use available materials and construct an arrangement of each box. Mark the price. Mark one box at $2. Mark another box at $3.

Sort the crayons in each set by color.

Are the colors of one set preferable to the other?

What choices do you have $(3 + 3 + 2; 2 + 2 + 2 + 2)$? Why not one $3 and the others $2?

Which choices do you prefer?

B. A science teacher wants to buy some new equipment. She has a total of $100 to spend. Test tubes cost $.50 each. Beakers cost $.10 more than twice the cost of a test tube. Test-tube holders cost half as much as beakers. Slides

FIGURE 10.5

can be bought for $5 per box (24 to a box), and a package of worksheets can be bought for one-half the cost of a box of slides.

The teacher needs 24 of each item except the test tubes, for which the need is 72. Any money left over can be used to purchase other materials. Any money that is not used will be given to the history teacher.

How can we help the teacher decide what to buy?

Plan: Display the information. Prepare a chart of individual costs.

Determine the total costs.

Determine what is left.

Determine what is most important (e.g., extra test tubes because they get broken).

Determine the cost of additional items.

Determine the amount not used.

Determine what the teacher can do if she goes over $100.

The two problems are essentially the same. The quantitative components vary in complexity, and the topics vary in level of sophistication. The crayon problem might be used with children in the second or third grade, whereas the science problem would be appropriate at the upper grade levels. Each problem provides for originality and flexibility. There are ample opportunities for negotiation, substitution, and exchanges. Opinion is important.

2. A problem is given. Alternative solutions need to be evaluated.

The school library does not have enough books and reference materials. Some concerns are:

A. If we ask for more money our taxes will increase.

B. If we ask for more money, the finance committee might delete all the library funds.

C. If we ask for more money, the finance committee might only approve part of the new amount.

D. If we ask for more money than we need, the finance committee might approve all of it. Then what?

E. If we do not get the materials, we cannot complete our assignments. What can we do?

Groups could be organized to determine the amount of money that is needed.

Groups could write up lists to justify the requests.

A group could play the role of the finance committee. Another group could play the role of the teachers. They could determine what will happen under each condition.

Votes can be taken to establish priorities.

Find out what approach the children prefer. Structure the groups to include handicapped and nonhandicapped children and give each person a role consistent with his or her present level of functioning. Conduct the activities.

3. A problem is given. All solutions are ambiguous.

A suggestion has been made that the number *two* be removed from our number system.

What problems will this cause?

How can we get along without it?

What alternatives do we have?

The teacher might elicit responses to each of the questions.

These requests can be written on the chalkboard. Different children can present the pros and cons of each response.

Ask the children to describe the role of the number *two* in our present number system.

What function does it have?

The class could vote (rank) the alternatives in order of their benefits or liabilities.

The teacher would stress the open-endedness of this situation and explain why it is hypothetical.

4. A problem is to be created. Solutions are needed.

This type of activity falls under the heading of brainstorming. Children are encouraged to be creative and to see if they can develop a problem that will be of interest to others. In the following example, children make up the problem.

Examples might resemble the following:

A. Pretend you are on a space ship and that you will soon land on another planet. Your job is to teach arithmetic in a new school. There are no books or materials. Much worse, they do not have the same number system you do. What can you do?

B. Pretend that you have just learned about measurement. You know both the English and metric systems. People do not like either. You have been asked to create a new system and the materials that will go with it. How will you do this?

Children would have to choose between the two examples. A key factor would be for the teacher to seek out each child's understanding of each problem. Children could set time limits and work in groups to identify their alternatives and solutions to topics such as those illustrated above. A key point is to create argument on a basis of preferences and other criteria. Encourage differences, but ascertain their soundness. Have children point out errors in logic or accuracy.

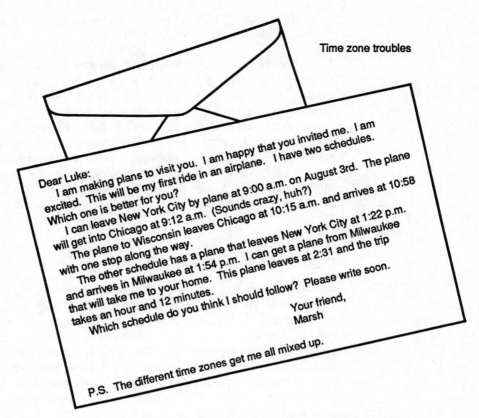

Time zone troubles

Dear Luke:
    I am making plans to visit you. I am happy that you invited me. I am excited. This will be my first ride in an airplane. I have two schedules. Which one is better for you?
    I can leave New York City by plane at 9:00 a.m. on August 3rd. The plane will get into Chicago at 9:12 a.m. (Sounds crazy, huh?)
    The plane to Wisconsin leaves Chicago at 10:15 a.m. and arrives at 10:58 with one stop along the way.
    The other schedule has a plane that leaves New York City at 1:22 p.m. and arrives in Milwaukee at 1:54 p.m. I can get a plane from Milwaukee that will take me to your home. This plane leaves at 2:31 and the trip takes an hour and 12 minutes.
    Which schedule do you think I should follow? Please write soon.

Your friend,
Marsh

P.S. The different time zones get me all mixed up.

1. How long would the trip be for Marsh if he took the morning schedule?

2. How long would the trip be for Marsh if he took the afternoon schedule?

3. How long is the flight from Chicago to Luke's home?

4. How long is the flight from Milwaukee to Luke's home?

5. How long does Marsh have to wait in Chicago?

6. How long does Marsh have to wait in Milwaukee?

7. How long is the flight from New York to Chicago?

8. If the flight from New York to Milwaukee takes an hour and 32 minutes, why does Marsh's schedule show only 32 minutes?

9. Which schedule would you follow? Why?

FIGURE 10.6   **Relating language arts to quantitative tasks**

5. Story production
Much of the literature relative to writing skills among the mildly handicapped indicates that the children manifest a variety of weaknesses or shortcomings. Writing exercises, whether they be geared toward production or editing, can be integrated into mathematics experiences.
A. Production and Editing
Organize the class into two or more groups. Direct the children in one group to write a story that begins with this sentence:

> Three astronauts will lift off in the space ship *Columbia* for a scientific mission.

Direct the children in the other group to write a story that ends with this sentence:

It is March 17 and three astronauts have just landed the space ship *Columbia* after 244 hours in space.

Note that one group will write a story that flows from beginning to end, whereas the other group must write to the conclusion. The stories of the members of each group could be assessed for mathematical and scientific accuracy as well as against standard writing criteria.

One or two stories could be modified by the teacher and given to the children to edit. Certain content and skill errors could be planted in the story and the children graded according to their correction of the story.

Children could be asked to write stories, brief statements for newspapers, and letters. Figure 10.6 shows a letter activity in which the focus is time and time zones. Any topic can serve as the core of the activity.

DECISIONS AND ARGUMENTS

Many quantitatively based activities possess a unique potential to provide experiences in organizational skills, scheduling, sequencing, information gathering, data collection, and program modification. One result of these efforts is to assist the individual to develop decision-making skills. Another result is to provide the individual with the bases for developing arguments and for evaluating an argument in a logical manner. Three types of activities—situations, applications, and social-personal units of study—are used to present experiences with decision making and argument.

Situations

Figures 10.7 and 10.8 contain worksheet activities that focus on decision making.

The example of Don and Dana illustrates a situation in which two master mechanics need to plan their schedules. The stress is on time-event relationships. The learner needs to determine the approximate amount of time it takes to perform each task, so there exists a need to interview, to collect data, to collate the data and to organize it into an acceptable schedule. Learners could be directed to different garages; they could obtain estimates and make comparisons. A need to be flexible is introduced when Mrs. Anselmo arrives with an emergency problem with her carburetor.

A number of basic skills are required in this lesson. The learner must be able to add and subtract and to perform these operations on denominate numbers. The learner must be able to take time estimates and fit them into a sequence that will enable each mechanic to work the same number of hours, even though one may do more jobs than another. If the learners are sufficiently sophisticated, the concept of flat-rate work could be introduced and examined.

The second illustration (Figure 10.8) does not require arithmetical computation. All data are provided in the chart. The necessary skill involves reading and matching numbers. The focus is on the relationships in the parts of triangles of different measure. There is no formal mention of the terms of the triangle (e.g., base), although informally this is a focal point in the lesson. Subsequent lessons could develop the vocabulary and the mathematical skill.

The situation in this second activity involves the placement of certain firefighting equipment so as to make the best fit between the building on fire and the equipment that will be most suited for that building. The task could be complicated by stipulating conditions such as time of day (e.g., 5:20 P.M.

Don and Dana were the master mechanics at Republic Garage. It was Tuesday and they had a lot of work to do. They wanted to plan a schedule. Dana took all the work orders and sat down with Don to plan the day. These are the work orders they had.

Job	Estimated time
Kevin's '74 Vega—tune-up, lube, oil, filter	
Mr. Toob's '79 Malibu—lube, oil, filter, rotate tires	
Anny B's '71 VW—generator problems	
Ms. Stearn's '78 Cadillac—fix door latch	
Dwayne's '76 truck—brake job on all four wheels	
Dr. Sikland's '79 Special—install CB, lube, oil, filter	
Freddie's '71 Bird—redo disc brakes, replace shocks	
Mary B's '79 2-door—lube, oil, filter, align front wheels, balance wheels	
Marcia's 4-door—service ignition, replace switch in blinker light	
Pappy's '79 truck—lube, oil, filter, fix electric windshield wipers	
Louise's Ford—replace front tires, check ignition, fix windshield wipers	

Make a chart like the one below. Fill in the schedule for Dana and Don.

Job	Don		Dana	
	Time start	Time finish	Time start	Time finish
1				
2				
3				
4				
5				
6				
7				
8				
9				
10				
11				

At 11:30 A.M., Ms. Anselmo came in with an emergency. Her carburetor needed replacement. How could she be worked in? Where?

FIGURE 10.7   Don and Dana: daily schedules

when the traffic is likely to be heavy), a given amount of time for each learner to complete the task so the fire can be put out as soon as possible, or by making available a map of the downtown area and requiring each youngster to name the streets. Children could be sent out to locate an actual fire station, collect data on average response time, the number of and types of fires, and examples of how the members of the fire department would solve the problem. Children could be encouraged to defend their plans.

Applications

Applications represent a type of activity in which the skills and concepts of one or more areas are applied to real-life experiences. Special education has long revered applications in all its curriculum areas. However, in many instances, the activities developed as applications were far removed from any real-life circumstance. Foremost among these are the proverbial tax return,

FIVE ALARM FIRE

A bad fire had started downtown. Five buildings were already on fire. Seven fire trucks had arrived on the scene. More fire trucks were on the way. The fire chief had many problems. One problem he had to figure out was how to get each fire truck as near to the buildings as he could. He also had to figure out how high the ladder would reach on each building. The fire chief made a chart to help him. His chart looked like this:

	Length of Ladder	Height of Building	Distance from Building
Truck A	39'	36'	15'
Truck B	40'	24'	32'
Truck C	13'	12'	5'
Truck D	18'	16'	10'
Truck E	37'	35'	12'
Truck F	34'	30'	16'
Truck G	12'	9'	3'

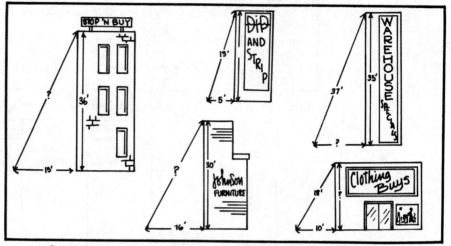

1. Which truck should the chief sent to "Stop and Buy?"
2. The chief sent truck D to "Clothing Buys". Was this correct? What is the height of the building of "Clothing Buys?"
3. The chief wanted to put truck E at "Warehouse Specialties." The truck could get as close as 10'. Would this be correct?
4. Which truck should be sent to "Dip and Strip?"
5. How long a ladder did the chief need for Johnson Furniture?
6. Which trucks were not used by the chief?

FIGURE 10.8 Problem solving with geometry

the reconciliation of the checkbook, and the payment of rent. These activities were almost always carried out as paper-pencil activities in a workbook.

Special education faces a new challenge with regard to applications. Where there are self-contained classes and self-contained curricula, the opportunity to include applications in the classroom is accentuated. However, when the individuals are enrolled in regular classes, particularly at the secondary level, the demands of meeting state or school requirements, the need to pass the competency test, the need to be proficient in math skills for vocational education, and the need to do the math that other kids are doing reduces the opportunities for applications. As McLeod and Armstrong (1982) have shown, the special education resource teachers tend to be extensively involved in helping the child to keep up with regular class assignments.

Special education seems to have faltered at three points. First, there have been too few attempts to systematically incorporate an emphasis on applications throughout the school years. Second, experiences in what was designated as real-life applications consisted of in-seat workbook activities. Third, with the advent of regular-education competency-test programs and the in-

Use this newspaper ad to answer questions 8, 9, and 10.

```
The Latest Model
EXCELLENCE
FULLY EQUIPPED FROM THE FACTORY
$90.86 per month
Full Price $4331
25% Down
Payments are for 42 months
at A.P.R. of 9.82
Deferred Payment Price $4899.29
```

8. What does the abbreviation A.P.R. mean in this ad?
   A. Average price reduction
   B. Advertised price review
   C. Annual percentage rate
   D. Annual protection rate
9. How much money is needed for a down payment on the Excellence?
   A. $25.80
   B. $90.86
   C. $119.62
   D. Not given
10. If Sam buys the Excellence on the deferred payment plan, how much does he pay in interest?
   A. $568.29
   B. $4331.00
   C. $4899.29
   D. Not given

FIGURE 10.9   Example from experimental version of minimum essentials test

volvement of mildly handicapped children in the test programs or the classes that were preparing children to take these tests, special education has not maintained its "real-life" emphasis. The back-to-basics movement is a movement away from applications.

Real-life applications should be as real as possible. In a visit to a special education class in Atlanta, it was observed that the teacher had developed a cardboard simulation of a shopping center. The center contained a bank and shops of different types. Youngsters actually played roles and simulated the actions and conditions of the shopping center. They took money to the bank, cashed checks, got money in return, went to the shops, and made purchases. In a Louisiana program, the youngsters go to the shops and make purchases from vending machines or from the store counter.

Figure 10.9 displays an item from the pilot version of the *Minimum Essentials Test* (Rice et al. 1980). Notice that this task requires specific knowledge (APR) or an ability to think logically to determine the meaning of APR. It requires skill with percent (e.g., convert 25% to .25 to multiply) and with the subtraction of whole numbers and decimals. The task also requires knowledge of the term *Deferred Payment Price* as the sum of the cost plus the interact and recognition that the difference between *Full Price* and *Deferred Payment Price* is the amount of interest paid.

Notice also that the response options include a *Not given* for two of the items. A youngster whose computations produces a response other than one of those shown should pick the *Not given*. However, many tests do not contain a *None of these,* and the youngster who has actually computed incorrectly is forced to pick one of those given. A test-wise child will attempt to match his or her response with one of those given. Other children may mark just any option so that they can go to the next item.

As we view programming from the metacognitive or strategy perspective, test-wisedness should be an integral consideration. At the very least, the child with a response that does not match one of the choices should be trained enough to redo the actual computation.

Social-Personal Concerns

Mildly handicapped children manifest a variety of social-personal traits that limit their full participation. These range from difficulties in the interpretation and meaning of social situations to individual problems such as failing to follow directions. Given our perspective that mathematics has to do more for the mildly handicapped child than focus solely on the academic qualities of mathematics, social-personal activities are integral to our programming efforts (Cawley et al. 1974, 1976). The mechanism for this focus is the Social Utilization Unit (SUU).

Figure 10.10 contains a SUU from Level 4 of *Project MATH*. In all, there are 50 units in Levels 3 and 4. Time requirements for the completion of the SUU's range from a day to a few months. If a school utilized the SUU's as the basis for a junior-senior high school curriculum, it is unlikely that all 50 units would be completed in the six-year span. Each SUU can be repeated.

The SUU's facilitate social-personal development through the use of individual and group experiences. The SUU's stress a discovery or inductive approach to problem solving and utilize interpersonal relationships, management of self to fulfill individual responsibilities, and short- and long-term task completion requirements.

Each SUU contains four major components. These are (1) Social Emphases, (2) Quantitative Concepts, (3) Necessary Skills, and (4) Unit Content.

SUU: Transportation Between Home and Job
Length: Two weeks

SOCIAL EMPHASES
1. Locating and acquiring information regarding public transportation schedules and costs
2. Making decisions based upon data
3. Working with a group to compile data
4. Working independently to solve a problem
5. Deciding what effects distance, weather, cost, convenience, reliability, and time have on the selection of a form of transportation to get to and from work

QUANTITATIVE CONCEPTS
1. Difference in the cost of a one-way trip ticket and a round-trip ticket or purchasing tickets in multiple quantities
2. Monthly cost of different types of transporation
3. Time and convenience of different forms of transportation as a function of cost
4. The effect of weather and traffic conditions for determining departing and arriving time at work for different types of transportation

NECESSARY SKILLS
1. Simple money notation
2. Basic operations of adding, subtracting, and multiplying
3. Simple map construction
4. Reading numbers

INTRODUCTION
This SUU may be introduced by one or both of the following activities.
(1) Discuss the need to get to work on time and also to get home from work at various hours if you are working shifts. Discuss the different ways you could get to and from work. Discuss how the locale in which you live determines the number of options available to you to get to and from work. (A rural area may have few forms of public transportation; an urban area, many.)

FIGURE 10.10 continues

(2) Make a list of all the possible ways you could get to and from work. Be sure to include such options as walking, car pools, paying someone to drive you, and different forms of public transportation.

PROCEDURE I

Procure a map of your city, county, or district, and put it up in the room. Make separate cards of different jobs and possible locations that can be found on the map—factory worker, Acme Industries; nurse's aid, City Hospital; etc. Make separate cards for different work shifts—8:30 A.M./4:30 P.M., 3:00 A.M./11:00 A.M. Make at least one of each kind of card for each learner. Give one job card and one shift card to each learner. Have the learners find their homes and jobs on the map. From the list of different transportation forms completed earlier, select as many as three different options for getting to and from work for these jobs and locations. If options involve public transportation, find out where to get information on bus, subway, and train routes, schedules, and prices; then pick up the information. Determine whether you would have to transfer to a different transportation route or a different method of transportation. Find out where to transfer and the times involved. If options involve personal transportation (car, motorcycle, car pool), get information on insurance costs, gasoline costs, parking fees, etc. Summarize all this information in a table such as this one:

Transportation	Estimated time going to work	Estimated time returning from work	Average cost for round trip	To arrive at job by _____ I must leave home by:
1. _____	_____	_____	_____	_____
2. _____	_____	_____	_____	_____
3. _____	_____	_____	_____	_____

PROBE QUESTIONS

What kinds of transportation are possible? Which one is best for the particular job and your home? What other alternatives are there? If your job is done at night, is taking a bus or subway possible? Do these means of transportation operate all night, or do they stop at a certain time? Are monthly rates possible for some of these methods of transportation? If you drive a car, what problems with parking might you have? Would it be very expensive? Would it be cheaper than taking a bus or subway, or more expensive? Should you allow sufficient time in case of traffic tie-ups? How does weather affect your leaving times and the type of transportation used? Is the cheapest form of transportation you found also the fastest and the most reliable?

EXTENSION

Elaborate on the preceding task by working in small groups and pooling all information. Summarize it into a table in order to answer the question, "Is there an ideal or perfect form of transportation, or do all forms have certain advantages and disadvantages?" Use the original list of transportation sources as the basis for the table. The summary *might* look something like this.

Trans-portation	Convenient	Reliable	Cost	Affected by weather?	Affected by traffic?	Can be used for all shifts?
Taxi	yes no	yes no	yes no	yes no	yes no	yes no
Car	yes no	yes no	yes no	yes no	yes no	yes no
Carpool	yes no	yes no	yes no	yes no	yes no	yes no
Bus	yes no	yes no	yes no	yes no	yes no	yes no

PROBE QUESTIONS

Which forms of transportation are considered low in cost by the group? Do all other groups agree? Do all groups agree as to which types of transportation are convenient or reliable? Did you leave any important things out of your table? Does the distance from your home to your place of work affect the decision as to which form of transportation is best?

FIGURE 10.10 Social utilization unit

The *Social Emphases* are a list of three to six social or personal traits that are the target behaviors for the unit. That is, in the unit shown in Figure 10.10, making decisions based on data, working with a group to compile data, and working independently to solve a problem are behaviors that are focused on in the unit. The activities of the unit include specific instances where the learner must perform these behaviors. The teacher manages and responds to these from whatever classroom management perspective she or he works from. If the teacher works from an ego-supportive system such as that of Redl (1959) and Morse (1980), the teacher might conduct an informal life-space interview and talk about the problem with the child. If the teacher works from a behavioral model such as that advocated by Hewett (1968), the teacher would observe the child and structure the learning experience to elicit the desired behaviors. Each teacher would reinforce the desired behaviors as they emerge.

A second component of the SUU is *Quantitative Concepts.* These are broad generalizations about the effects or meanings of quantity. In the present illustration, cost differences between one-way and round-trip ticket purchases is one of the concepts highlighted.

Necessary Skills are those math skills that are needed in the group to complete the tasks of the unit. Not every child needs to have every skill. This is managed by assigning children to certain roles. Not every learner has to perform every task. Some might be able to multiply, others only add. Because the SUU is conducted as a group activity, members of the group can be assigned different responsibilities. A child who can prepare a copy or original of a simple map can be given that responsibility. A child who can add can be given responsibilities requiring that skill.

The practice of role differentiation enables the teacher to focus on the needs of an individual. If the child needs an opportunity to work with others, the role for that child would be developed accordingly. Youngsters can be grouped together in a manner that is mutually supportive.

The content of the SUU is described in two or three pages. It is presented in outline form to allow the teacher a substantial degree of freedom in making adaptations to local situations. The *Probe Questions* direct the attention of the learner to selected concepts. They are intended to be informal in that the teacher might bring them up during a discussion of behavior or data.

APPRAISAL

The term *appraisal* is defined to mean a comprehensive review of the child from developmental and diagnostic perspectives (Cawley 1985b). The developmental approach seeks to determine present levels of functioning as they relate to curriculum components (i.e., the math problem in the books or teacher tests) or normed references such as a standardized test. The diagnostic approach seeks to determine the difficulties unique to an individual on a specific set or type of problem.

To say that the problem-solving components of our curricula and the components of our appraisal systems are less than adequately developed would be a gross understatement. Our earlier citation of Schoenfeld's (1982) observation that better than 90 percent of the word problems in one elementary series could be solved by use of cue words is representative of the shortcomings in curricula. Our previous illustrations of the inconsistencies in *Key MATH* are illustrative of the appraisal shortcomings.

Effective appraisal in problem solving is predicated on the same principles that guide appraisal in computation or concept development. Among these considerations are:

1. Problem-solving appraisal is contextual or situational. The tactics one uses should vary with the context in which the appraisal takes place.
2. Appraisal should be multimodal; it should include manipulatives, group and individual tasks, and it should be capable of identifying the effects of one variable on another (e.g., if child cannot read the material he or she may not get to the problem; reading should be adapted to level of child).
3. Appraisal should minimize the gaps between different types of problems and include an acceptable representation of the processes and characteristics of the problems.

Effective appraisal is contextual or situational. One would use one approach to the appraisal of word problems that systematically vary problem traits (e.g., include or exclude extraneous information). Another approach would be used in the administration of a test where the computational priority controlled the problems, but the problems themselves were not defined in any way; examples of such problems would include single-digit by single-digit items at the beginning and three-digit by three-digit items at the end. One clearly needs a scheme that differentiates correct and incorrect responses as a result of computational factors from one that differentiates correct and incorrect responses by choice of operation. One would appraise performance differently from the creations perspective than from the applications perspective. The format of appraisal when using the Social Utilization Units has its own unique qualities.

Matrix Appraisal

Developmental appraisal. A reasonable approach to the development of curricula and appraisal is matrix programming (Cawley et al. 1979b; Goodstein 1974). Matrix programming begins with the designation of the components of a matrix or a set of matrices. Figures 10.11 and 10.12 present two different matrices.

Matrix programming enables one to organize curriculum or appraisal to produce materials with reasonably defined sets of characteristics. Matrix programming is decision-oriented. Each assignment given to a child is based on her or his performance on the previous assignment. This contrasts markedly with an approach in which the next assignment is simply the next page.

The components of the matrix in Figure 10.11 are as follows:

1. Reading levels: first, second, third
2. Information processing: extraneous information/no extraneous information
3. Language structure: simple sentence/simple sentence with indirect object
4. Language vocabulary: simple subject/simple object
5. Computation: addition of single-digit by single-digit with sums 9 or fewer, to three-digit by two-digit, no renaming

The following are illustrative of the problems that would emanate from Figure 10.11.

No. 1: An Indian boy had 2 ponies.
Another Indian boy had 3 ponies.
How many ponies did the boys have in all?

	Reading vocabulary level						
	First		Second		Third		
	Non-extraneous	Extraneous	Non-extraneous	Extraneous	Non-extraneous	Extraneous	
Simple sentence	1	2	3	4	5	6	
Simple sentence with indirect object	7	8	9	10	11	12	Single digit by single digit
SS **SS w/IO**	13 19	14 20	15 21	16 22	17 23	18 24	Two digit by single digit, no renaming
SS **SS w/IO**	25 31	26 32	27 33	28 34	29 35	30 36	Two digit by two digit, no renaming
SS **SS w/IO**	37 43	38 44	39 45	40 46	41 47	42 48	Three digit by two digit, no renaming

SS—Simple sentence
SS w/IO—Simple sentence with indirect object

FIGURE 10.11 Matrix illustration

No. 8: An Indian boy gave his uncle 2 hides.
An Indian girl gave her uncle 3 hides.
Another Indian boy gave his uncle 4 hides.
How many hides did the boys give to the uncle?

No. 41: An Indian tribe captured 326 buffalo.
Another Indian tribe captured 61 buffalo.
How many buffalo did the tribes capture?

No. 48: An Indian tribe with many ponies captured 326 buffalo.
An Indian tribe with few ponies captured 63 buffalo.
Another Indian tribe with many ponies captured 62 buffalo.
How many buffalo were captured by the Indian tribes who had many ponies to start with?

	First reader vocabulary				
	Direct addition	Indirect addition	Direct subtraction	Indirect subtraction	
CS/SO	1	2	3	4	Single digit by single digit
SS/CO	5	6	7	8	
CS/CO	9	10	11	12	
CS/SO	13	14	15	16	Two digit by two digit
SS/CO	17	18	19	20	
CS/CO	21	22	23	24	

FIGURE 10.12 Matrix illustration (CS = complex subject; CO = complex object; SO = simple object; SS = simple sentence)

Planning the assignments for each child requires that the instructor be familiar with the characteristics of the matrix. The instructor makes the assignment by designating certain problems for each child as shown below:

Child 1: Assigned to do 1,2,3,4,5, and 6.
What factors did teacher vary?

_____

_____

Child 2: Assigned to 1,2,7,8,13,14,19, and 20.
What factors did teacher vary?

_____

_____

Child 3: Assigned to do 1,7,13,19,31, and 43.
What factors did teacher vary?

_____

_____

On the second day the instructor assigned Child 1 items 12, 18, 24, 30, and 36. Assuming the child had all problems correct on the first day, what was the teacher doing with this child?

_____

_____

As can be seen, the instructor provided many individual adaptations. Not all the children did the same number of problems. Not all the children did the same type of problems. The teacher was able to adapt to these individual needs because matrix programming is quick and reasonably precise. The characteristics of the problems are specified so the instructor knows what was assigned to each child. Assume someone asked for an explanation of the different assignments. The instructor might explain as follows:

Child 1: I had this child do problems in the first row because I wanted to check him on reading vocabulary level and problems with extraneous information that were all written in very simple sentences. After I found out he could read the third-reader problems and do them correctly, I sent him down the column to see if he would have any difficulty with another sentence and other types of computation.

Child 2: I wanted to see how this child would do with both sentence formats and items with and without extraneous information when I kept the vocabulary level at the first-reader stage. I changed the computation.

Child 3: This is a child who reads at the first-grade level and does not do well with problems that contain extraneous information. I wanted to give him problems that would take advantage of his good computation.

The matrix in Figure 10.12 limits itself to first-reader vocabulary and to single-digit and two-digit computation. It is likely that the problems of this matrix will be more difficult than those in 10.11 due to the inclusion of direct and indirect problems.

Illustrative of the problems that would emanate from this matrix are:

No. 1: A boy had 3 apples.
Another boy had 2 apples.
How many apples do these boys have?

No. 2: A boy added 3 apples to his pile.
He now has 4 apples.
How many apples did he start with?

No. 15: A girl had 16 apples.
A boy bought 14 of her apples.
How many apples does the girl have left?

No. 16: A girl has 12 apples left after she gave 13 to a boy. How many apples did the girl start with?

Utilizing the matrix to make assignments, a teacher might give the following:

Child 1: Assignment was to do 1,2,3 and 4.
What factors did the teacher vary?

_____

_____

Child 2: Assignment was to do 1,5,9 and 13.
What factors did the teacher vary?

_____

_____

Child 3: Assignment was to do 2 and 4.
What factors did the teacher vary?

_____

_____

By referring to the matrix, the essential characteristics of each assignment can be tracked. The teacher is able to explain the differences between the set of tasks given to each child and to validate the individualism of each. The purpose of each matrix is to get each child to the cell in the lower right hand corner as rapidly as possibly. Once proficiency in this cell is attained, the child would move to a different matrix. Children who can move rapidly through a matrix would not be burdened with a need to do every problem. Children with difficulties could be taken through various combinations that would meet their needs. The teacher can prepare additional problems instantaneously by matching the characteristics of a cell with the needs of a child.

Diagnostic appraisal. Within the present context, the term *diagnosis* refers to an in-depth analysis of an individual learner on a single or limited type of problem solving. The diagnostic process is designed for use with a child who is experiencing inordinate problems (Cawley 1985b).

The in-depth analysis of the child requires the following:

1. There should be a clear delineation of the content difficulty. That is, one should determine the type of problem that poses the difficulty. This could be accomplished by having the child complete sample problems in different matrices. It is as important that the examiner determine what the child can do as it is to determine what the child cannot do.
2. The impact of related skills should be determined. If the youngster cannot read the problem, the difficulty could be reading level, not problem solving. As is the case with appraisal and instruction in computation and concepts, we advocate a multimodal approach to problem

solving. The Interactive Unit (IU) provides a system to guide one through an alternative-modality approach to diagnosis.

3. The teacher and the child should interact in an interview format. The teacher might ask the child to build a representation of a problem or to act out the problem. The teacher might build a representation of a problem and ask the child to state or write in words the problem the teacher is presenting.

4. The child should be given a set of four or five problems of the same characteristics. The teacher can ask the child to take some objects and to build a representation of each problem. The teacher needs to observe the reliability of the strategy the child uses, the efficiency, and the sequence in which the child approaches each problem. A child might do one in an efficient and correct manner and then get confused on another. A child might do them all wrong using the same strategy. Of even greater concern to the teacher is the child who does each one wrong in a different manner. In the case of the last child, it is suggested that the teacher go to a more familiar set of problems and have the child build on strengths.

Diagnosis is a clinical process that is likely to be as individual to a teacher or examiner as are the differences among children. It is suggested that the steps used in diagnosis be written down so that they can be backtracked when an unusual case emerges.

HOW DO WE TEACH PROBLEM SOLVING?

We teach problem solving by teaching it regularly and systematically throughout the twelve to fourteen years the child is in school.

We teach problem solving as a set of processes and behaviors that can be systematically enhanced by the use of good materials.

We teach problem solving by freeing the child from any difficulties in getting to the problem. If reading or computational difficulties impair problem solving, they should be modified. A basic difficulty with traditional problem-solving activities is their all-inclusive character. That is, as one element of the problem increases, so do all others. The problem in a sixth-grade book is more difficult, the reading level is more difficult, and the amount of work is greater. In effect, the traditional program expects the child to develop simultaneously in all areas of proficiency. By definition, this is not the developmental nature of learning disabilities.

We teach problem solving by creating contextual and situational settings in which children can solve problems and understand that they are solving problems. We accentuate their awareness to problems, guide or direct them in the process of solving problems, and then mandate that they demonstrate the proficiency independently. Given a familiarity with a variety of problem traits such as described in this chapter, the teacher can ad-lib or make up problems at any time. The approach can be subtle and informal, or it can be direct and for a specific purpose. Children need to see that their efforts were of value and that some closure was obtained.

We teach problem solving from an interactive perspective. That is, acting out or creating representations of problems and their solutions has as much validity as paper-pencil worksheets, possibly even more. There are trade-offs in every system. One trade-off in an interactive system, such as we propose, is that some children experience as much if not more difficulty when switching from one modality to another. For example, in one study (Goodstein & Sedlak

1974) it was found that mildly retarded children remembered the information in orally stated, class-inclusion problems more quickly than learning-disabled children. However, once the learning disabled were provided with enough repetitions so they remembered the information, they got substantially higher scores.

We teach problem solving by assisting children to understand the nuances of vocabulary and language structure. We do not develop one-to-one correspondence between a word and an operation. We help the children to generalize by embedding words in different structures and by calling attention to these variations.

ASSESSMENT FOR MATHEMATICS INSTRUCTION

INTRODUCTION

Assessment of achievement or aptitude in mathematics has as its purpose more effective and more efficient teaching efforts. Such assessment either leads to or flows from mathematics instruction. A model summarizing the roles of different kinds of assessment in the instruction process is illustrated in Figure 11.1.

Assessment before instruction meets one of two purposes: placement or diagnosis. Post-testing is usually criterion referenced, norm referenced, or domain referenced. In Figure 11.1 the space around the positions of these various types of assessment represents the necessary connectedness among them. In this chapter we will examine the nature of these evaluation procedures and establish criteria to be considered when assessment instruments in mathematics are selected or developed.

PLACEMENT TESTING

Perhaps the most salient characteristic of placement testing is its potential for effecting a proper match between a student and an instructional program. A program may be a commercially produced curriculum, a syllabus developed by a local school system, one developed for a local school system, or both of the latter. Usually such a program is developmental in nature, rather than remedial. With excess in neither the number of items nor the time necessary for administration, the placement instrument must serve as an entrance into a program at a point appropriate for the student involved. Usually placement tests are criterion referenced.

Most criterion-referenced placement tests are built around a sampling of the behavioral objectives upon which the related program is built or a selection of topics from the mathematics program into which the student is to be placed. Such instruments are often organized by level (e.g., grade level) or by content area (e.g., geometry.) The items on a placement test should reflect a balance between concept assessment and skill measurement.

At times, in order to circumvent the poor match between achievement level and the chronological age of a student, a norm-referenced mathematics

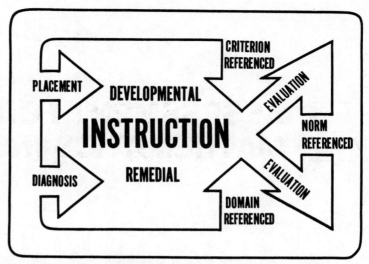

FIGURE 11.1   Relationships among kinds of assessment and instruction

instrument is administered as a *placement test*. The subsequent *grade-equivalent score* is then used as a basis for placing the student at a particular level in a mathematics program. Dangers inherent in such an approach will receive greater attention at a later point in this chapter.

Because of the likelihood of different patterns of performance even among students who seem to achieve at identical grade levels, placement tests based on the actual content of a curriculum are the better choice. Such instruments offer another advantage: the relationship between the test and content of the program is such that what was "placement" testing prior to the beginning of the program can be used as "exit" testing after a sequence of instruction, and, in cyclic fashion, such "exit" testing can function as "placement" testing for the next sequence of instruction. Viewed in this fashion, placement testing becomes an integral and ongoing component of a mathematics program.

Some placement tests reflect a one-to-one correspondence between test items and items of content. For some children the results of such a test may not be very reliable. For these children the format of the *Kraner Preschool Math Inventory* (Kraner 1976) may provide a solution. This test is designed so that

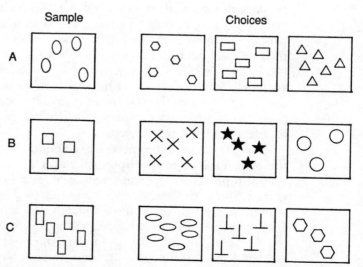

FIGURE 11.2   Kraner Preschool Math Inventory: items for set equivalence

Chapter Eleven

there are three items per content topic. A student does at least two of the items. If the results do not agree, the student is asked to do the third item.

For example, a child's ability to identify a set equivalent to a given set might be tapped by the first two items in Figure 11.2. A child who succeeded on both items *A* and *B* and the child who succeeded on neither *A* nor *B* would progress to another task. A child who succeeded on one item and failed on the other would be asked to do the optional item. His or her performance on this third item would determine a score of success or failure for that topic.

Although Kraner does not seem to address this possibility, a record of the number of optional items given to a child would seem to provide some diagnostic information on its own. The child who achieved a high score and answered correctly the first two items for each topic would seem to be different from the child who achieved the same high score and required the optional items frequently.

DIAGNOSTIC TESTING

The purpose of diagnostic testing in mathematics programming is different from that of placement testing. Many, if not most, mildly handicapped children are capable of learning mathematics in a developmental program, provided the program is structured to meet the needs of such children with respect to content, pace, and methodology. For some students, because initial instruction has been faulty, or because other experiences or handicaps have intervened, placement in a development program is not appropriate. A closer look at the patterns of achievement and learning of such youngsters is necessary. For such students diagnostic testing is indicated if worthwhile instruction is to take place.

As the diagram in Figure 11.3 illustrates, diagnosis concerns itself with the combinations of what content (concepts and skills) a student has acquired, what combinations of teacher-learner interaction are most appropriate for that student, and what rules a student has developed for himself or herself as a result of previous experiences with mathematics. Instruments accomplishing less should be regarded as making a restricted and limited contribution to the achievement of the purpose of diagnosis.

Content Analysis

The determination of what mathematics a mildly handicapped student knows may be the easiest task in the diagnostic process. Inventories of knowl-

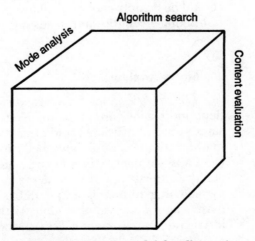

FIGURE 11.3   A model for diagnosis

edge of mathematics are numerous. They range from standardized achievement tests to inventories serving as placement tests for individual programs or even for individual classroom grouping. Of the former, perhaps *Key Math, SRA Diagnosis: An Aid to Instruction,* the Brigance instruments, and the WRAT Arithmetic test are among the best known. Any of these or similar instruments can provide an estimate of what mathematics a student knows.

Some factors influencing the choice of such instruments demand attention. To avoid an unduly long testing session, the diagnostician should focus on the areas of mathematics content that comprise a set of knowledge reasonably expected of the student being tested. For a young child, the operations of arithmetic beyond simple addition would be excluded from the content analysis in order to provide time in which the examiner can determine the degree of knowledge of other areas of mathematics; for example, basic geometry concepts or problem-solving strategies. For a secondary student who has in the past demonstrated some knowledge of basic concepts of number, a reasonable starting point in the content analysis might be the operations of arithmetic. In other words, the examiner, after selecting a measure of mathematics content, should refine the instrument by administering only those portions of the test thought suitable for a given student.

A second factor that should influence the choice of tests providing the content analysis is the opportunity afforded to examine a student's knowledge of different areas of mathematics. Minimally this set of content should include number concepts (whole number and the other rational numbers), basic arithmetic, basic geometry concepts, measurement skills, and problem solving. The investigation of a student's skills in each of those areas will provide a pattern of strengths and weaknesses. Such a pattern serves as an aid in the development of a remedial program for the student under consideration.

A third consideration governing the choice of a test of content is the ease with which the instrument can be adapted to circumvent any difficulties provided by the format of the test itself. A test that is solely paper and pencil might intimidate a student who has had many unpleasant experiences with such instruments. A test that requires skills in reading beyond those already acquired by a student may fail to reveal what the student actually knows about mathematics. Some students have not acquired certain test-taking behaviors— for example, matching the items in Column A with those in Column B, selecting the correct answer from four choices, or writing answers in blank spaces found at the right hand side of the page. Such students may be unnecessarily hampered in their efforts to express what is known.

Content analysis, then, is the first step in the diagnostic effort. Ideally, the instrument(s) chosen to accomplish this task would be efficient and should test knowledge in several areas of mathematics. In addition, the test format should be flexible enough to allow an examiner to adapt the instrument to the test-taking strengths and needs of a student.

Mode Analysis

The investigation of what modes of instructional interaction between student and teacher are most suitable for a given student is necessary from the standpoints of efficiency and effectiveness. Very often, students referred for mathematics diagnosis have been in school for a number of years. Other (perhaps younger) students are referred because a deficit in mathematics ability is already apparent. These students may experience extreme difficulty in matching numerals and number names, or in writing numerals, or in mastering number combinations or the like. In either case, previous forms of instruction have been ineffective. Mode analysis will point out which combinations can result in more effective teaching. At the same time, because such students have lost time, efficient instruction is necessary if they are to learn what needs to be grasped during the remainder of their formal academic experience.

Learning styles and the need to match instruction to an individual's preferred mode have captured the attention of educators during the past several years. Most efforts have reflected concern with modality, not learning. The auditory learner, for example, was described as someone who learned better when the input was "to the ear." Little was said about the student's output during the instructional event. Yet it is that combination of input and output that demands attention.

The following story illustrates the need for this type of investigation. A school psychologist was working with a moderately retarded student. The student had been referred for testing prior to placement in a vocational education program. The examiner included a mode analysis based on the Interactive Unit (see Chapter 2) in the testing procedure. After going through the sixteen input/output combinations a few times, the examiner noted that the combination of say input and manipulate output proved to be very difficult for the student. That observation was particularly important in that case, because the say/manipulate interaction was one that would have occurred frequently in the vocational education program of the student.

The Interactive Unit provides a structure for the input/output analysis. Used in the context of diagnosis, the Interactive Unit serves to provide some

M/M	D/M	S/M	W/M
Examiner builds model of triangle. Child does the same thing.	Examiner points to diagram of triangle. Child builds model of figure shown in diagram.	Examiner defines/describes triangle. Child builds model of triangle.	Examiner shows *triangle* written on card. Child builds model showing what the word means.
M/I	**D/I**	**S/I**	**W/I**
Examiner builds model of triangle. Child selects picture of what was built from among three choices: square, rectangle, triangle.	Examiner points to diagram of triangle. Child selects same diagram from among several choices.	Examiner defines/describes triangle. Child selects diagram of what was defined/described from among several choices.	Examiner shows *triangle* written on card. Child selects diagram showing what the word means from among several choices.
M/S	**D/S**	**S/S**	**W/S**
Examiner builds model of triangle. Child describes what was built or names the figure.	Examiner displays diagram of triangle. Child describes the figure shown in the diagram or names the figure shown.	Examiner defines/describes triangle. Child repeats definition/description.	Examiner shows *triangle* written on card. Child reads the word aloud and/or defines/describes the figure.
M/W	**D/W**	**S/W**	**W/W**
Examiner builds model of triangle. Child draws picture showing what was built, writes word triangle, or selects word from three choices.	Examiner displays diagram of triangle. Child copies the figure, writes the word naming the figure, or selects the word from among several choices.	Examiner defines/describes triangle. Child writes definition/description or draws picture showing what the definition/description means.	Examiner shows *triangle* written on card. Child copies the word, writes a definition/description, or draws figure showing what word means.

FIGURE 11.4 The Interactive Unit and the content triangle

indication of which combinations work best for a given student. By selecting a bit of content with which the learner is familiar and developing sixteen activities around that content, the examiner can work with the student to determine which combinations provide the best learning opportunities for the student.

Repeating the procedure for a new piece of content for which the learner has the necessary prerequisites can provide further information that will help in the development of a mathematics program for that student. In Figure 11.4 we present an illustration of the sixteen combinations built around the topic triangle. An examiner who had carried out the sixteen activities with a student who knew triangle prior to the testing session might proceed to repeat the sixteen activities around a shape unfamiliar to the learner—trapezoid, for example. Such a procedure provides valuable information about the behavioral combinations that are most effective and efficient for that student.

The reader will note that none of the sixteen activities is lengthy in nature. The entire sequence can be undergone in a relatively short space of time. The mode analysis provides an efficient way to determine how best to instruct a student.

The Algorithm Search

This third component of diagnosis has as its purpose the searching out of the rules of operations a mildly handicapped student has developed. Usually rules of computation are the primary concern, but a learner's strategies in other areas may also be of some interest; for example, counting strategies, tactics used to determine certain measures such as perimeter and area, and the like. This section will be confined to algorithms in the context of arithmetic computation.

Examine the examples provided in Figure 11.5. Most teachers are familiar with one or more of the computational procedures illustrated there. More often than not, students practice the behaviors until mastery and habituation occur.

Some of the rules illustrated represent inappropriate strategies that almost never result in a correct answer. Of greater danger are algorithms like those shown in examples (e) through (g). In example (e), the strategy of "carrying" only in the column at the farthest left works when that is the only column in which regrouping is required. Dividing from right to left is appropriate only when there are no "remainders" from the second step—unless, as

a.		b.	c.	d.
27		27		32
31		34	432	× 4
+62		+ 62	−117	78
12		1113	325	(4 × 2 = 8, 4 + 3 = 7)
(2 + 7 + 3 + 1 + 6 + 2)				

e.	f.	g.
	1	
1	1	
237	247	$81r2$
152	153	$5\sqrt{425}$
+360	+362	
749	852	(5 ÷ 5 = 1, 42 ÷ 5 = 8)

FIGURE 11.5 Incorrect algorithms

in example (g) of the illustration, the student recognizes the remainder as 20 and writes it as 20/5, a most unlikely event. These algorithms are particularly hazardous because they work sometimes; the student is reinforced periodically for what is basically a faulty tactic.

In many cases, the reasoning behind the adoption of certain faulty rules seems apparent. Students spend many hours in the primary grades adding single-digit numbers in columns and in rows. When faced with example (a) in the illustration, they maintain the single-digit addition behavior. In the case of subtraction, students may see the process as one of finding the difference between two numbers in a column rather than lessening the top number by the amount represented by the bottom number. Multiplication is often taught as repeated addition, and when there are two digits in the multiplier, the usual practice is to add the partial products. One of these approaches may explain the combination of addition and multiplication in example (c). The operations of addition, subtraction, and multiplication are usually performed from right to left. The student following the rule illustrated by the division example may have failed to make the switch from right to left to left to right. Ginsburg (1986, 107) reminded us that "like their accurate answers, children's errors are often produced by idiosyncratic but meaningful strategies." Mildly handicapped children often do follow rules of common sense in developing faulty algorithms.

Ferreting out nonconventional and perhaps faulty algorithms can be undertaken in two ways. The teacher should closely examine the written work of the student and attempt to figure out what the student has done. The teacher should also sit with the student and observe how the student completes an example on paper. When the student completes the work, the teacher asks the student to explain his or her procedure. The instructor can use a checklist of more common errors to check off faulty procedures and can write notes to record any unusual results. Because of the extra information it affords, the interview should be an integral part of the algorithm search. For the operations with whole numbers, the Buswell-John Diagnostic Chart provides a step-by-step procedure for this type of algorithm search.

Ideally, the algorithm search should be conducted by the person responsible for the mathematics instruction of the student. Many of the decisions about the remedial effort should be based on the results of the algorithm search. During the interview, the teacher can note the words and grammatical structures used by the student in the course of his or her explanations. These language characteristics can serve as a bridge for the efforts to remedy inappropriate computational behaviors and/or concepts.

Some students seem to have no consistent rules of computation, in spite of frequent exposure to such rules over a number of years. At first glance one might suppose that the remedial effort could be carried out more easily with such a student, since relatively little unlearning has to take place. Yet this lack of acquired habits should be a cause for concern.

The failure to pick up even faulty rules over a period of time suggests that such students may have serious problems rooted in noninstructional sources. For these students the mode analysis discussed earlier in this chapter is particularly important. Upon discovering that the student has acquired few, if any, computational rules, the instructor or examiner should return to the mode analysis for the purpose of further testing the student's responses to written input and the student's capabilities for responding by way of written output. Where serious deficits are found, recourse to other tools for computation—for example, a hand held calculator—would seem to be justified.

Items for a checklist to be used during the interview procedure can be gathered from the teacher's own experience or from reports detailing specific computational behaviors. Ashlock (1982), Buswell and John (1925), Reisman

(1978), Reisman and Kauffman (1980), and Cawley (1985b) are just a few of the many sources of such information. The reader is encouraged to pursue any or all of these references.

The preceding discussion has focused on the components of content measurement, mode analysis, and algorithm search as separate entities of the diagnostic process. In the actual carrying out of the diagnosis, the three tasks are usually intermingled. Of most importance is the awareness that good diagnosis is the optimal tool for flexibility in remediation.

ACHIEVEMENT TESTING

Given a set of information to be learned, the following factors usually play a role in the determination of the success of the learning task: the number of items to be learned; the number of items answered correctly on a test; the number of items answered correctly by other students; the number of trials necessary to learn the material; and the time taken to learn the material. Not all factors are given equal importance, however. Those factors that are considered most seriously give a clue as to the type of post-testing undertaken.

Let us consider an example. Twenty students in a class are required to learn ten multiplication facts.

Case 1. The teacher assigns the ten facts on Monday. All twenty students will be tested on Friday. A portion of each day will be set aside for drill and practice. Each student's score will consist of the percent of correct responses on Friday's test.

If the teacher were to picture the results of Friday's test on a graph, the outcome might look like the graph in Figure 11.6. The graph serves to picture each learner's score in relationship to the percent-correct score of other learners. A number of students did well, achieving scores between 75 percent and 100 percent; a number of students failed, achieving scores less than 50 percent; the scores of the remaining students fell in between the two extremes. If the teacher assigned another set of facts to be learned by the next Friday, some students would approach the new task without having learned the old material. One suspects that these same students will achieve comparably low scores on the next test also. A learner's cumulative record would depict his or her percent-correct scores over time.

Case 2. The teacher assigns the ten facts on Monday. Each learner is to rehearse the facts each day until he or she can answer all of the examples correctly in each of two consecutive trials. A record will be kept of how many trials each learner takes to achieve the criterion each day. In this case the learner's individual charts might resemble those illustrated in Figure 11.7.

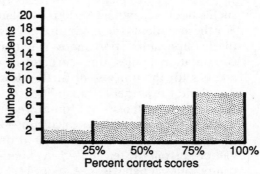

FIGURE 11.6 Percent correct

One learner's chart might look like this:

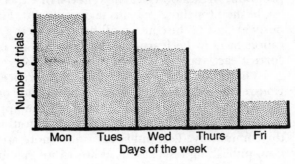

Another learner's chart might look like this:

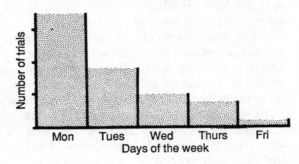

A third learner's chart might look like this:

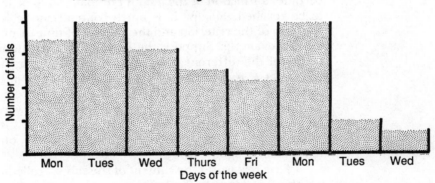

FIGURE 11.7   Trials to criterion

Notice that in this case the criterion of success each day was 100 percent mastery for all students. The variables to be measured for each student were number of trials each day and the number of days. Only after the mastery of the first set of examples would a student be allowed to start learning new material. Each student will improve his or her learning rate with each new set of material, so that over time a student's cumulative record will be one depicting rate of learning, rather than percent correct.

This latter case serves as a rough example of criterion-referenced testing; the former case illustrates briefly one of the principles behind norm-referenced testing. We will examine each of these more closely.

Criterion-Referenced Testing

Criterion-referenced tests have two characteristics. The items on such a test are usually derived from a set of instructional objectives. A set of criterion-referenced items usually reflect the content taught in an instructional program. In addition, certain criteria, or expected levels of performance, are established for those persons taking the test. For the example outlined previously in Case 2, the content of the items on the test was the set of ten multiplication facts to be learned. The criterion of performance was 100 percent correct each day for each of two trials.

The movement toward criterion-referenced testing, and hence to criterion-referenced instruction, has been a steady one over the past two and a half decades. School systems have adopted the approach in order to determine whether or not students are achieving educational goals and objectives set by the community. Test publishers show more and more progress in this direction, publishing tests based on lists of instructional objectives set by the community and reporting item-by-item results for each student tested. The developers of textbook programs in different areas such as math, science, history, and the like are advertising criterion-referenced evaluation instruments to accompany each program.

The implications of this movement for mildly handicapped children are several, especially when the subject matter is mathematics. Since criterion-referenced testing implies not only the existence of a set of content but also an established standard by which performance is judged, special attention must be paid to the second variable. The standard can be that of the number of trials to criterion or to the amount of time to criterion. Success in meeting criterion for a given set of content becomes a function of the time taken or allowed or the trials taken or allowed. The reverse is also true. An increase in the amount of content, or in the standard, or in both demands a longer time and a larger number of trials. Thus, the time or number of trials necessary becomes a function of the amount of material to be learned and the criterion to be reached. Finally, how much content can be learned to criterion is a function of the criterion and the amount of time or number of trials available.

An example: Suppose the body of content under consideration is the names of the different polygons, or P. Let t = the time taken or number of trials. Let C = criterion or standard of performance. If all of P are to be learned to a 95 percent value for C, then t will be a larger number than it would be if P were more restricted and/or if C were less than 95 percent. Likewise, if a learner had a large amount of t at his or her disposal and a relatively low value for C (say 60 percent), more polygon names might be a reasonable assignment.

In the real world the content of the subject called mathematics is limitless, but the amount of time a learner spends in school is very limited, and only a portion of that time is spent learning mathematics. Thus, some restraints must be placed on the body of mathematics content to be learned and, possibly, on the standard of performance to be reached. Our present concern is an examination of the advantages of criterion-referenced assessment for mildly handicapped students.

Mathematics has two characteristics that apply here. The first of these is the hierarchical structure of the system—one step leads to another, and most new knowledge depends on the acquisition of some prerequisites. The second factor stems from the role of mathematics in everyday functioning—few people progress through a day without frequent recourse to one branch of mathematics or another. The consideration of these two realities presents a strong argument for criterion-referenced assessment in the cause of mastery learning for mildly handicapped students. Through such assessment some limited guarantee that students achieve the knowledge of prerequisites and gain a degree of facility with everyday mathematics can be attained. The strength of that guarantee will depend to a large extent on the nature of the standard or criterion to be met.

Another consideration arises. Criterion-referenced instruction had as one of its origins the espousal by the educational community of such techniques as behavior modification, task analysis, and behavioral objectives. Each of these approaches to instruction is most easily implemented at less sophisticated levels of cognition, or to use the terms of Bloom (1956), at the levels of knowledge, comprehension, and application. Hence criterion-referenced instruments are most easily composed of items measuring those levels of learning. Care must be taken, especially in the area of mathematics, to develop items and instruments measuring other types of cognition. If instruction is directed toward the fostering of such skills as analysis, synthesis, or evaluation, test items will be more readily composed.

If instruction has been concerned not only with teaching students the formula for computing interest on a car loan (knowledge), but also with guidelines and procedures for comparing different options for a loan (knowledge plus evaluation), then items measuring both skills can be included in the achievement assessment. This type of instruction and assessment is particularly important for mildly handicapped children, because these students are not likely to pick up those more sophisticated skills on their own.

Norm-Referenced Testing

The premises underlying norm-referenced testing are quite different from those underlying criterion-referenced testing. For all students undergoing norm-referenced assessment, the body of content remains constant and need not be derived solely from an instructional program to which all students have been exposed. The question of time or trials to criterion becomes irrelevant. Norm-referenced assessment results in the placement of those taking the test along the continuum from severe failure to outstanding success. The location of mildly handicapped students is usually on the failure side; on that supposition the continuum was derived. That is to say, a large

Whole numbers

Item no.	Sam	Tina	Item no.	Sam	Tina
1			24		x
2			25		
3			26		
4			27		
5			28		
6			29		
7			30		
8			31		
9	x		32		x
10		x	33		
11			34		
12	x		35	x	x
13			36		x
14			37		
15			38		
16		x	39		x
17	x	x	40		
18			41		
19		x	42	x	x
20			43	x	x
21	x	x	44		
22	x	x	45	x	x
23	x				

FIGURE 11.8 Items answered incorrectly by two students on a norm-referenced mathematics achievement test on whole numbers

	Fractions	
Item no.	*Sam*	*Tina*
1		
2	x	
3	x	
4		x
5	x	x
6		
7		
8	x	
9	x	x
10	x	x
11	x	x
12	x	
13	x	x
14	x	x
15	x	x
16	x	
17	x	
18	x	x
19	x	x
20		x

FIGURE 11.9   Items answered incorrectly by two students on a norm-referenced mathematics achievement test on fractions

number of the items on a norm-referenced instrument are there precisely because they have distinguished the more capable students from those loss capable during the norming process.

A norm-referenced instrument serves to provide a comparison between the performance of a student and the performance of other students. The terms used to express such comparisons—for example, grade equivalent scores—can be misleading. A grade equivalent score of 8.3 in mathematics for a fifth-grade student does not necessarily mean that the fifth-grade student can do eighth-grade math. If such a score means anything at all, it says that the fifth grader does fifth-grade math as well as an eighth grader would. The reverse would also be true. A grade equivalent score of 2.3 for the fifth grader merely indicates that the student does fifth-grade arithmetic about as well as a second grader. Many test publishers who provide printed test scores for student records and parents are attempting to offer such explanations along with the test results.

Other types of scores reflecting norms post problems also. Chief among these is the tendency to equate the performance of students having approximately the same score (for example, stanine scores). The following examples are based on actual test results.

Sam and Tina were students in the same seventh-grade classroom. During the month of October the teacher administered the mathematics portion of a standardized achievement test. On the sections dealing with arithmetic computation (whole numbers and fractions) these two students achieved almost equal stanine scores. A look at the two students' performance, item by item, revealed the information presented in Figures 11.8 and 11.9. As the figures illustrate, the two students differed in their performance on eighteen items. To ignore these differences is to run the risk of failing to meet the needs of either Sam or Tina, or both.

Let us look at a second example. Dave and Ben were students in another school. During the same month they too took a standardized test. Dave's grade-equivalent score on the computation portion of the test was 4.2. Ben

Item no.	Dave	Ben	Item no.	Dave	Ben	Item no.	Dave	Ben
1			17		x	33	x	
2			18			34	x	
3			19			35	x	
4			20	x	x	36	x	
5			21	x		37	x	x
6		x	22	x	x	38	x	x
7			23	x	x	39	x	x
8			24	x		40	x	x
9			25			41	x	
10		x	26			42	x	
11			27			43	x	x
12			28	x		44	x	x
13		x	29			45	x	x
14		x	30			46	x	x
15			31	x	x	47	x	x
16	x	x	32	x	x	48	x	x

FIGURE 11.10   Items answered incorrectly by two students on a norm-referenced mathematics achievement test on whole numbers

scored 4.3. In Figure 11.10 are listed the incorrect items for each boy on the subtest on whole numbers. The boys' performance differed on 14 of the 48 items of the subtest. Of even greater importance is the fact that Dave indicated no answers for items 31–48. Ben selected an answer for each and every item. Once again, the test scores reveal neither of the above.

There are still other disadvantages to the use of norm-referenced achievement measures with mildly handicapped students. One of these is the result of the fact that the norms are supposed to mirror the performance of normal or relatively intact learners. Slow and therefore delayed development in one or more areas is characteristic of the mentally handicapped. The use of norms to measure student gain can actually mask actual growth.

One reason for this is the nature of norms as a measuring instrument. Some measuring scales are designed so that the numbers we use represent proportions. A two-pound box of sugar is twice as heavy as a one-pound box of sugar. A ten-pound bag of sugar is twice as heavy as a five-pound bag of sugar. A six-foot pole is three times as long as a two-foot pole. A fifteen-foot pole is three times as long as a five-foot pole. Norms do not act this way. A grade-equivalent score of 4.0 does not reflect twice as much knowledge as a grade equivalent score of 2.0. A stanine score of 6 does not reflect three times as much knowledge as a stanine score of 2.

Now suppose that a teacher were to use *Key MATH* as an achievement measure at the beginning of a school year and at the end of that same year. For a child of average capability, one would expect an achievement gain of one school year over a school year's time. If instruction in many different areas of mathematics had taken place, one could also reasonably expect comparable growth in all areas. In other words, if the child's overall grade-equivalent score on the pretest was 1.4, then at the end of the school year the child's performance should be around 2.4.

What is a reasonable expectancy for a mildly retarded child having an IQ of 65? If grade-equivalent scores were proportional in nature, one might expect a gain of 65% of a school year, or approximately six or seven months. However, grade-equivalent scores are not proportional. Hence it would be quite possible for a mildly retarded student's test score to represent little improvement over the course of a school year, even if instruction had been both comprehensive and effective. Likewise, given the right circumstances, a student's score could represent an inflated level of achievement.

If grade-equivalent scoring is necessary for administrative reasons, an approach like that of the *Brigance Diagnostic Inventory of Basic Skills* may serve the purpose more beneficially. The *Brigance* combines a criterion-referenced format with specific grade-level performances determined for each objective. In this way a specific instructional plan can be developed for a student, with grade-equivalent scores available to whatever agency deems such scores necessary.

Domain-Referenced Testing

The one-to-one match between instructional objective and test item, which is peculiar to criterion-referenced testing, often results in an unwieldy set of test items. In reality, more often than not, so called criterion-referenced tests supply not a one-to-one correspondence between item and objective, but rather a sampling of items, each of which represents or embraces a number of specific objectives. Such tests are more properly called domain-referenced tests. Each item on a domain-referenced test is derived from a domain of content. That domain may be the subject of several specific instructional objectives. Examine the following set of objectives:

Subtract a one-digit number from a two-digit number, regrouping tens

Subtract a two-digit number from a two-digit number, regrouping tens

Subtract a two-digit number from a three-digit number, regrouping tens

Subtract a two-digit number from a three-digit number, regrouping hundreds

Subtract a two-digit number from a three-digit number, regrouping tens and hundreds.

Rather than including five examples on a test, an example such as $213 - 67 =$ might be selected to represent the domain of content specified by the five objectives. If a student succeeds in getting the correct answer, one expects that he or she has achieved the four preceding objectives also. A failure to obtain the right answer probably means that one or more of the objectives under consideration has not been met.

Viewed from the perspective of practicality, domain-referenced tests have much to offer both the cause of efficiency and the need to provide for individual assessment. We suspect that domain referencing is the technique used by most classroom teachers trying to assess the effectiveness of instruction. When administered in such a way that the disabilities of a student would not handicap the demonstration of knowledge, these tests are appropriate for handicapped students. Students performing in a satisfactory manner can be placed at the appropriate level for the next instructional goal. The work of students who do not achieve at the expected level can be analyzed, and, if necessary, further testing (criterion-referenced in format) can be undertaken.

The emphasis on the link between assessment and instruction leads us to consider specifically the concept of comprehensive appraisal.

COMPREHENSIVE APPRAISAL

It has been proposed (Cawley and Miller 1986) that a comprehensive approach to appraisal in mathematics is fundamental to our ability to meet the needs of the mildly handicapped. In this context, a comprehensive appraisal consists of two general domains. The first domain provides for the appraisal of mathematics. The second domain provides for the appraisal of abilities and disabilities that affect performance in mathematics, but which are not a part of mathematics per se. Appraisal in either or both domains can be undertaken from curriculum-based or norm-referenced perspectives.

Within each of these domains and perspectives consideration must be given to the determination of instructional and independent levels of functioning. It is necessary to determine a level at which the child can be properly and efficiently instructed and a level at which the child can work independently in an effective and efficient manner.

Instructional Levels

The instructional level is the level at which the teacher and the learner are directly involved in the interaction between content, process, and time. The instructional level has been previously defined as the level at which the teacher teaches, such that the child does not make more mistakes than the teacher can properly correct (Cawley 1985b). What this means is that in situations in which the child makes an excessive number of errors and the teacher is unable to attend to these errors, the teacher is not at his or her instructional level. Please note that the emphasis is on defining the instructional level of the *teacher*, for the teacher controls the level of content, the selection of processes to be used to deliver the content, and the amount of time the teacher can direct toward the topic or to an individual child.

The instructional level of the child can be defined to consider the rate at which the child responds to items when they are first taught. Rate should be comparable to that of other learners at the same level. Number correct should always be some four to five times greater than number incorrect.

The instructional level needs to be determined in a manner that gives the child an opportunity for maximum level of functioning. If, for example, the child can do long division, but can read only at the first-grade level, it would not be appropriate to give the child long division activities when the instructional and other reading materials are written at the sixth-grade level. By the same token, it would not be proper to give the child first-grade math activities to accommodate the reading difficulty. What is appropriate is to select the proper level of mathematics content (e.g., long division) and then determine a set of processes through which the child can interact with that math (e.g., avoid reading and use oral presentation). This is easier said than done because the teacher has only a limited amount of time (e.g., 40 minutes to distribute among 25 children in a regular class and 40 minutes to distribute among 5 or 6 children in a resource center).

The Independent Level

The independent level is the level at which the child can perform without assistance. This suggests the child has the necessary prerequisites in reading, language, and cognitive development and that the level of content is appropriate. The independent level includes all individual tasks in the classroom and all out-of-class assignments such as homework. It is imperative that the child be able to function effectively and efficiently for the independent level. By this we mean the child should be able to get a substantial proportion of the items correct and that he or she should be able to accomplish the task in about the same amount of time it takes other children. If it takes the mildly handicapped child 3 hours to do a homework assignment that other children do in 30 minutes, the task is inappropriate for the independent level for it suggests something less than preparedness with the prerequisites or the content.

In establishing the independent and instructional levels, there needs to be a distinction between learning and achievement comparable to that in Chapter 1. Time, number of repetitions, amount recalled in both short- and long-term tasks and a reduction in the time or number of repetitions on subsequent encounters with the same material are factors that relate more to learning than achievement. These same factors are the ones that enable the teacher to plan instructional activities more effectively than content selection alone.

One differentiation between instructional and independent levels of functioning (Cawley 1985b) stressed the need to consider high levels of pupil performance at the entry and exit stages of either level. Unlike reading, where numerous authorities have proposed performance criteria for instructional and independent levels, mathematics educators have yet to propose comparable standards. Our own work of the past twenty years suggests the following to be reasonable as a beginning set of standards:

Independent level

Possesses all prerequisites in reading, language, and cognitive ability
Be able to read and understand all directions and relate them to the present task
Computation: Entry-level standards
    Accurate at the 95% level within operation
    Accurate at the 90% level across operations
    Rate and performance comparable to others
Problem Solving: Entry-level standards
    Accurate at 99% level with operations
    Accurate at 95% level with reading
    Accurate at 95% level to set up problem
    Accurate at 70% level at entry stage
    Accurate at 95% level at exit stage

Instructional Level

Computation: Entry-level standards
    Accurate at 90% level on prerequisite operations/skills
    Accurate at 0% when first taught
    Rate and time for acquisition comparable to others
    Accurate at 79% level after completion of first instructional sequence
    Accurate at 90% level after completion of total instructional sequence
    Accurate at 75% level to express operation and concepts with alternative representations (e.g., manipulatives)
Problem Solving: Entry-level standards
    Accurate at 99% level with operations
    Accurate at 95% level with reading
    Accurate at 0% set up when first taught
    Rate and time for acquisition comparable to others
    Accurate at 70% level at setting up problem representation after first instructional sequence
    Accurate at 90% level after completion of total instructional sequence
    Accurate at 70% level for problems of different types across operations

Differentiations between instructional and independent levels of functioning would be such that until the child attained criteria at the instructional level no assignments would be given at the independent level. Naturally, there has to be a distinction between practice exercises such as completion of a workbook assignment in class for the purpose of reinforcement or teacher evaluation and those assignments that are clearly independent.

Curriculum-based or subject-matter-based appraisal focuses on comparisons of a given learner with others in a similar or the same setting. These types of appraisal utilize the books and other instructional materials of the classroom as the primary content referent. Often, the tasks (for example, paper-pencil routines) are the same. This has the advantage of demonstrating to the teacher just how the child is functioning within the classroom. The mathematics teacher often knows how the child functions because assignments and teacher-made tests are used frequently.

A partner in the appraisal process is the norm-referenced or standardized test. Norm-referenced tests have two major weaknesses in addition to those described previously. First, they allow content to be determined statistically. That is, the items selected for the final version are those that meet certain statistical criteria such as differentiating among samples in item analysis. This results in content gaps and a less than adequate sampling of mathematics. The *Key Math Diagnostic Arithmetic Test* is an example of a test that has so many content gaps that its results are meaningless for instructional or placement purposes. It would be far better to develop a set of content and then determine the statistical qualities of the items that represent that content. A second weakness of norm-referenced tests is that they lack comprehensiveness. By this we mean they do not appraise the individual across a sufficient range of content or process.

Cawley and Miller (1986) have developed a scheme for a comprehensive mathematics appraisal. This consists of the following three components.

Measurement. This contains three norm-referenced tests of computation, listening vocabulary, and reasoning or problem solving. Each component has been designed from a content perspective. That is, the items in the computation test are sequenced according to their characteristics. Use is made of similar items. For example, 42×6 in the numbers section is repeated as 4.2×6 in the decimal section to provide a more direct measure of the skill. Each item in the reasoning section is defined according to its characteristics. If appraisal of the role of extraneous information is desired, it is included in a specific item.

Assessment. This is a five-part instrument that focuses on school-related, but not necessarily school-learned, mathematics in basic concepts, numbers, fractions, geometry, and measurement. Figure 11.11 shows one item from each of the subtests. The items are designed to range from preschool through the equivalent of the upper middle grades and in most instances include nearly all the math content that substantial numbers of mildly handicapped children will master.

Diagnosis. The diagnostic component consists of an intense analysis of individual performance within a single operation. That is, a child determined to be in need of diagnosis in a single operation such as division would be intensively studied within division. Ordinarily, this child is more seriously disabled in this one area than in others. The diagnostic component is more fully described in *Practical Mathematics Appraisal of the Learning Disabled* (Cawley 1985b).

SUMMARY

In the preceding pages, we have examined some of the specifics of assessment and children with handicaps to learning and achievement. One basic principle

Basic concepts
Classification by number

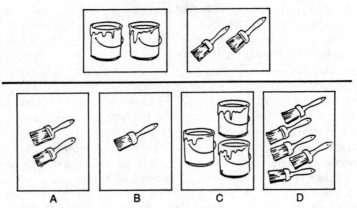

Look at the pictures at the top. From the choices below the line,
point to the one that has the same number of objects as those shown.

Numeration
Set recognition

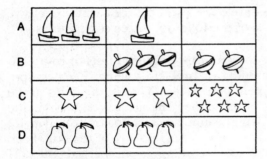

Point to each row of pictures and say:
  a. Point to the set with only one.
  b. Point to the set with none.
  c. Point to many.
  d. Point to the picture with no pears.

Fractions
Part-whole relations

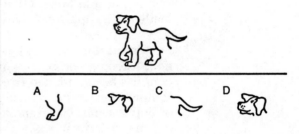

Point to the dog above the line and say:
  Look at the dog at the top. Point to the picture below the
  line which shows the part of the dog that is missing.

Geometry
Parallelism

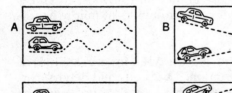

Look at the pictures of the cars and the roads they will
travel. Which picture shows that the two cars will meet?

Measurement
Weight

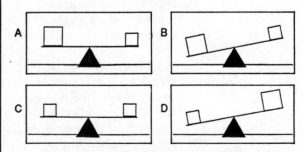

Look at the pictures. Which one shows that the small
box is heaviest?

FIGURE 11.11 Sample assessment items

228

underlies such attempts at testing in the area of mathematics: All assessment leads to or flows from instruction. Pretesting can take the form of placement testing, usually based on instructional objectives, and diagnostic testing. The latter assessment should embrace the components of content, mode analysis, and algorithms.

Post-testing formats are usually criterion-referenced, domain-referenced, or norm-referenced in nature. Both types of instruments imply certain outcomes for mildly handicapped students. In particular, scores obtained from norm-referenced instruments must be interpreted very cautiously, if instructional needs are to be met.

Most authorities argue strongly that tests meet certain statistical standards and that they be culturally and socially responsive to variations in learner experience and background. It is still beyond the comprehension of some of us that it took a court order and/or a set of federal regulations to tell us that it was inappropriate to test children in one language when they were dominant in another. However, we continue to recognize a need for greater sensitivity to human variability when appraising mildly handicapped children. For this reason, we advocate the use of comprehensive procedures that include multi-modal appraisal, the differentiation between learning and achievement, and a representation of content in mathematics that is independent of what the child might have learned only in school.

We recognize the need to produce instruments that meet the various statistical qualities that make them reliable and valid. At the same time, we believe that the production of instruments that are highly reliable but do not yield equivalent magnitudes of reliability in interpretation are of less than full value to the mildly handicapped. The ultimate validity of an appraisal for the mildly handicapped is attained when the appraisal provides a sufficient enough description of the child to provide a basis for accurate curriculum placement and for the determination of specific approaches to instruction.

MULTIMEDIA MATERIALS: INSTRUCTIONAL AIDS IN TEACHING MATHEMATICS

INTRODUCTION

Multimedia materials or instructional aids used in the teaching of mathematics serve to multiply the teaching-learning options, especially for individuals with characteristics that hinder their development of mathematical concepts and skills. The heart of any curriculum is the teaching-learning process, which is composed of the strands of (1) learners (with their many characteristics and abilities), (2) content, and (3) methodology. Objectives establish the relationships between learners and content, and various content-related strategies join content to methodology. Significant contributors to the strategies are the materials that are the teaching tools that enhance instructional efforts. Within a multiple-option curriculum for the mildly handicapped learner, such multisensory materials are very important if we are to achieve all of the interactives of instruction.

For the purposes of this chapter, *instructional aids* are defined as any materials or devices other than a basic mathematics textbook, which a teacher may use in the instructional process. The aids may be used by an entire class, a small group of learners, or an individual. These aids may be used in the lesson presentation, as study aids, or for enrichment purposes. The following outline demonstrates the range of instructional aids.

INSTRUCTIONAL AIDS

Printed material
Content outlines for unit or chapter
List of unit objectives
Workbooks
Worksheets
Programmed booklets
Review sheets
Study aids

Activity cards
   Drill-and-practice cards
   Problem-solving cards
   Enrichment cards
Diagrams and pictures
Charts
Tables
Graphs
   Pictograms
   Bar graphs
   Line graphs
   Circle graphs

Basic equipment
Straightedges
Stencils for shapes
Scissors
Compasses
Rulers and tape measures
Protractors
Clocks
Thermometers
Volume and capacity containers
Scales and weights

Models and manipulatives
Demonstration boards (felt)
Place-value devices
Attribute blocks
Pattern blocks
Number boards
Geoboards
Measuring devices
Kits
Games and puzzles
Science apparatus
Charts
Plane and solid geometry models

Electronic devices
Tape recorders
Slide- and filmstrip projectors
Film projectors
Videocassette recorders
Overhead projectors
Calculators

Computers

Laser disk equipment

It is *not* the purpose of this chapter to explore each element from this list; other more extensive publications have done this (NCTM 1973). In this chapter we will select various instructional aids, suggest a use for each aid, and provide selected examples within the context of this book. The major purposes of this approach are to expand instructional options, to provide ideas to motivate learners, and thus, to provide meaningful instruction for mildly handicapped learners. Since we live in a fast-paced world with instant communication and information explosion, we must use any sensory devices we can to compete with the world outside the classroom.

ACTIVITY CARDS—DEVICES FOR INDIVIDUALIZATION

With the renewed emphasis on problem-solving strategies and the individual nature of the actual process of problem solving, activity cards that branch from a given content strand can serve many functions while being simple instructional aids. An activity card contains the description or diagram of a series of steps to complete or finish a task. Since many such cards are teacher-made and somewhat apart from the regular curriculum, they serve to motivate and reward learners. The same activity can be represented in different ways, and the activities may be individual or small-group activities.

ACTIVITY

FROM SKIP COUNTING TO SQUARE NUMBERS

Mathematics has at times been defined as a search for patterns. When students learn to count, "1, 2, 3, 4, 5, . . . ," they are working with a number pattern or a series. Skip counting of, "1, 3, 5, 7, 9, . . . ," and "2, 4, 6, 8, 10, . . . ," also exemplify this approach. Using this continuous process and geometric figures, we can produce a series of square numbers on activity cards.

The first card contains the directions to arrange four stickers in a pattern so that if you draw lines from sticker to sticker you produce a square shape. On the back of the card represent the answer (see Figure 12.1).

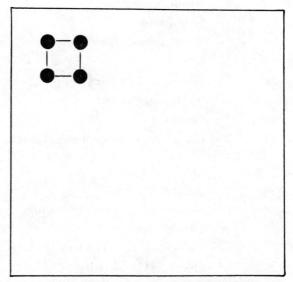

FIGURE 12.1 A square with four dots

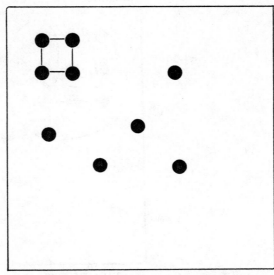

FIGURE 12.2 A square and five random dots

The second activity card contains a square shape of four stickers (in our diagram, dots) and five more dots with the instructions to make a larger square using the first (four-dot) square and the five additional dots (see Figure 12.2).

We can assist the learners to establish a pattern (see Figure 12.3).

At this point we have several options with respect to the third card. Some learners may be asked to look at the first two solutions and either tell how many dots would be needed for the next largest square or be asked to construct such a square with stickers. Other learners may need to be guided to construct the sixteen-dot square. Still other learners may be able to extend the idea to a twenty-five-dot square or even beyond.

The pattern produced is 4, 9, 16, 25, 36, . . .

Laboratory Activities as an Approach to Teaching

Other activity cards may be arranged to produce a series of laboratory activities that function together in a series or independently to lead to the discovery of a concept or relationship:

One set of laboratory activities can be used to develop the idea of a number relationship. One instruction card may instruct the learners to collect circular objects, such as a quarter, a jar lid, a hoop, and the like. Using a piece of string and a ruler, the learners will be asked to determine the distance around the given figure (circumference) and the longest distance across the figure (the diameter). On a chart (see Figure 12.4), the learners are asked to record their findings, to add the two measurements, to subtract the smallest from the largest in each case, to multiply them, *and* to divide the smallest into the largest. (Note that these operations place these activities within the curriculum after some understanding of division.)

The student is led to discover that the division process produces similar answers even though the measures are different; thus, we have a discovery of the relationship that produces the number pi (π).

At higher grade levels we could introduce probability activities and calculator activities (with a series of numbers) to produce independent activity cards that lead to pi if steps are followed in the experiments.

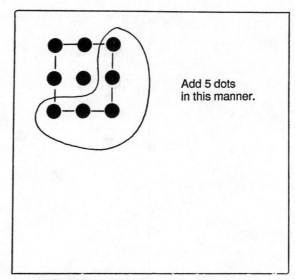

Add 5 dots
in this manner.

FIGURE 12.3   A square with nine dots

A Problem on an Activity Card

Individual problems on activity cards provide excellent opportunities to meet individual needs and provide enrichment activities. We need many options here, since a problem for some individuals may not be a problem for others. Our definition of a problem is a situation to which the solution is not readily apparent to the potential problem solver. The following problem is an example of activity-card presentation and demonstrates that trial and error is a part of problem solving in the form of *tentative hypothesis.*

Problem situation

We have a box 2 feet long, 2 feet wide, and 2 feet high. Under these dimensions, the surface area is 24 square feet (2 × 2 × 6). The box needs to be wrapped in order to mail it. The only piece of wrapping paper is 3 feet by 8 feet or 24 square feet.

Working at nome, the stock person discovered a way to cut the piece of wrapping paper into two pieces and cover the box by taping the two pieces together.

Use a piece of paper 3 inches by 8 inches and cut this paper into two pieces so

Measuring circular objects

Object	Distance around (C)	Distance across (D)	C + D	C − D	C × D	C ÷ D

FIGURE 12.4   Measurement table

Chapter Twelve

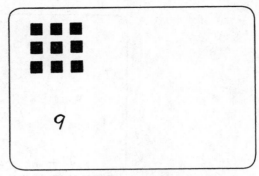

FIGURE 12.5   Nine units

that you can fold the two pieces, tape them together, and form a box 2 inches by 2 inches on each side.

    If you can complete this activity, then you can do the big problem. (The answer is shown in Figure 12.16 at the end of this chapter.)

USING THE OVERHEAD PROJECTOR TO GO FROM THE OBJECT MODE TO THE PICTORIAL MODE OF INSTRUCTION

Among the accepted teaching tools in the mathematics classroom, the overhead projector is a good choice for use in demonstrating and building a mathematical concept. To demonstrate its usefulness let's examine the mathematical concept of place value. On the glass surface we can demonstrate and review whole-number addition of basic facts. To introduce place value consider nine (9) and one (1) more.

Teacher input	*Learner output*
Place nine blocks of the same color on a piece of clear acetate on the surface of the overhead projector and ask a learner to write the number symbol for the number of blocks (see Figure 12.5).	A learner writes *9*.
Place one block of the same color as before beside the first nine and ask a learner to write the numeral for the number of the new blocks (see Figure 12.6).	A learner writes *1*.

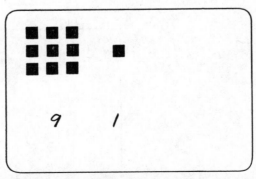

FIGURE 12.6   Nine units and one more unit

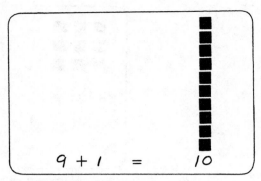

FIGURE 12.7   Ten units (ones)

Teacher input   *Learner output*

Move all blocks together to the right of
   where they were. Announce that now we
   have ten (10) blocks. Write the *10*. Write
   the *plus* and *equal* symbols (see Figure
   12.7).

Move the blocks together and tell the
   learners that the 10 units or blocks have
   now become 1 ten-block. Complete the
   process by trading the 10 units for 1
   ten-block. (See Figure 12.8.)

Give the learners a worksheet made by
   tracing the blocks that you used on the
   overhead.

Ask the learners to place blocks on the
   small squares and trade the ten blocks for
   1 ten-block. (See Figure 12.9.)   Learner trades 10 ones
                                                      for 1 ten and 0 ones.

Repeat this for each family of ten (10) (8 +
   2, 7 + 3, 6 + 4, 5 + 5, 4 + 6, 3 + 7, 2 +
   8, 1 + 9, 0 + 10, 10 + 0).

We started with objects. The projected image became a picture that we
duplicated on the worksheet; thus, the step from the object mode to a pictorial
mode has been reduced. In this example we have some guidelines for the
development of worksheets. Our first worksheets should be based on real
objects or models of real objects that can be used with the worksheets to test
what we are doing.

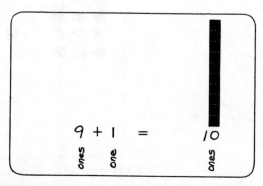

FIGURE 12.8   Trading ten ones for one group of ten

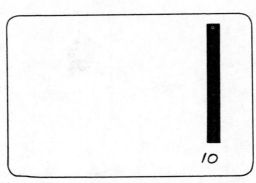

FIGURE 12.9   One group of ten

The overhead projector is the best device to use to demonstrate the trading process for regrouping. An example involving subtraction of fractions can extend this concept.

Teacher input	*Learner output*
Suppose we want to subtract ¼ from ½. Use your fraction kit and show ½ of a circular region.	
	Learner selects ½ region and places it on the worksheet.
Demonstrate on the overhead projector (see Figure 12.10).	
Use your fraction kit and show ¼ of your circular region.	
	Learner selects ¼ of same circular region.
To use the fraction kit for subtraction we need to change the ½ region. What shall I trade for the ½ region?	
	Learners respond by saying, "²⁄₄".
Demonstrate the trading process (see Figure 12.11) and ask the learners to do the same as on the worksheet.	
	Learners trade ½ for ²⁄₄ and place on worksheet.

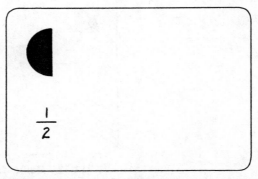

FIGURE 12.10   One-half of a circular region

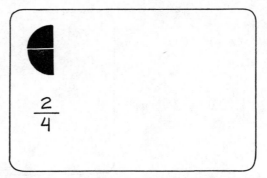

FIGURE 12.11   Trading one-half for two-fourths

Teacher input	*Learner output*
Now to show subtraction move a ¼ region to the right and label this answer.	
	Learners take ¼ region from ²⁄₄ regions and write answer on worksheet.
Demonstrate the subtraction process. (See Figure 12.12.)	
Now place the remaining ¼ region to the right and label completed exercise. (See Figure 12.13.) Ask the learners to do what you did.	
	Learners complete the subtraction and place correct symbols on the worksheet.

By introducing the process on the overhead projector we have an aid to use in the future if learners experience difficulty. The aid has been a part of the developmental process.

The overhead projector can also be used to aid in problem solving. One of the common activities used with tangrams is to ask the learner to take each of the seven pieces and arrange these pieces so that they form a square region. An instructor can form the square region on the glass surface of the overhead projector and slide the pieces apart to provide clues to the learners as they attempt to obtain a solution (see Figures 12.14 and 12.15).

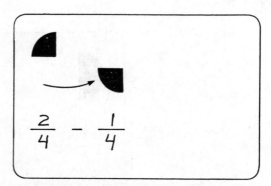

FIGURE 12.12   Removing (subtracting) one-fourth

The overhead projector can also be used to relate mathematics to other subjects, such as science. Using iron filings and a set of magnets, the instructor could demonstrate a magnetic field.

Despite its usefulness, however, the overhead projector should not be used as a replacement for the chalkboard.

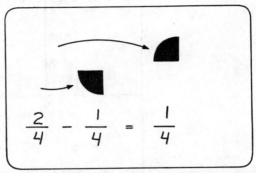

FIGURE 12.13   Moving remaining part (obtaining the answer)

FIGURE 12.14   Square region formed by tangram pieces

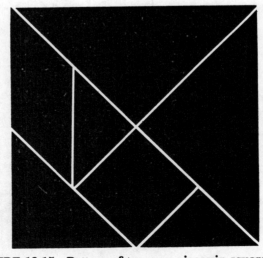

FIGURE 12.15   Pattern of tangram pieces in square shape

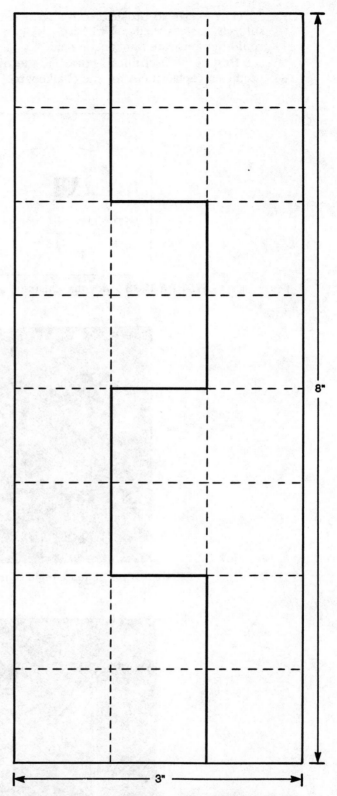

FIGURE 12.16   Solution to the problem situation

A MICROCOMPUTER AS A TEACHING TOOL

Now that we have emerged from the first wave of microcomputers and the emphasis on teaching programming, we can turn our attention to using a microcomputer as a teaching tool. We do not need to know everything about how a computer works in order to use it. Many people, for example, drive automobiles without knowing everything about engines. In the same way, the microcomputer should serve as a valuable tool even for those who do not fully understand computer technology.

Teacher utility programs that can be run by selecting an item from a menu on a disk or tape or by copying a given program for a given microcomputer from a magazine should be of use to everyone. Instead of redrawing a diagram for a ditto, a program can be saved on a disk and printed whenever the teacher needs the diagram as a handout. An example of this use of a microcomputer in mathematics is the development and use of a calendar.

```
5    REM THIS PROGRAM PRINTS OUT A BLANK CALENDAR FORM
10   HTAB 5
15   PRINT " :  SUN  :  MON  :  TUE  :  WED  :  THU  :  FRI  :  SAT  :"
20   FOR J = 1 TO 6
25   HTAB 5
30   PRINT " :_____:_____:_____:_____:_____:_____:_____:"
35   HTAB 5
40   PRINT " :       :       :       :       :       :       :       :"
45   HTAB 5
50   PRINT " :       :       :       :       :       :       :       :"
55   NEXT J
60   HTAB 5
65   PRINT " :_____:_____:_____:_____:_____:_____:_____:"
]RUN
```

SUN	MON	TUE	WED	THU	FRI	SAT

Other examples of this nature include forms for thermometers, scales, clocks, and geometric shapes.

Using programs such as the following, drill and practice sessions may prove to be a motivating influence in the classroom, especially if graphics are introduced to respond to the maturity level of the learners. This program in

BASIC gives the learners (or their instructor) the following options: operation (addition, subtraction, multiplication, or division), a number of examples, exercises, problems, and a choice between easy or more difficult problems. The program also keeps a record of right and wrong answers.

```
10    PRINT "THIS IS A DRILL AND PRACTICE PROGRAM"
11    PRINT "IN MATHEMATICS"
15    PRINT "YOU WILL BE ABLE TO CHOOSE EITHER"
20    PRINT  TAB( 10) "ADDITION"
25    PRINT  TAB( 10) "SUBTRACTION"
30    PRINT  TAB( 10) "MULTIPLICATION"
40    PRINT  TAB( 10) "DIVISION"
50    PRINT "YOU WILL BE ABLE TO SELECT THE NUMBER"
51    PRINT "OF PROBLEMS AND WHETHER YOU WANT EASY"
52    PRINT "OR MORE DIFFICULT PROBLEMS"
55    PRINT
60    PRINT "WHAT IS YOUR NAME? ": INPUT N$
65    PRINT "HELLO, "N$", PLEASE SELECT AN
      OPERATION"
70    PRINT  TAB( 10) "ADDITION-TYPE A"
75    PRINT  TAB( 10) "SUBTRACTION-TYPE S"
80    PRINT  TAB( 10) "MULTIPLICATION-TYPE M"
85    PRINT  TAB( 10) "DIVISION-TYPE D"
95    REM  INPUT CHOICE OF OPERATION
100   INPUT C$
110   IF C$ = "A" THEN GOSUB 200
120   IF C$ = "S" THEN GOSUB 400
130   IF C$ = "M" THEN GOSUB 600
140   IF C$ = "D" THEN GOSUB 800
150   PRINT
160   PRINT "WOULD YOU LIKE TO CHOOSE A DIFFERENT"
161   PRINT "OPERATION? (Y OR N)": INPUT Q$
170   IF Q$ = "Y" THEN GOTO 70
175   PRINT "ARE YOU SURE YOU DON'T WANT MORE"
176   PRINT "PRACTICE? (Y OR N)": INPUT B$
180   IF B$ = "N" THEN GOTO 160
185   PRINT "I'VE ENJOYED WORKING WITH YOU, "N$
190   PRINT "HAVE A GREAT DAY"
195   END
200   REM PROGRAM WILL RUN ADDITION PROBLEMS
205   REM
210   PRINT "SELECT THE NUMBER OF PROBLEMS YOU
      WOULD"
211   PRINT "LIKE TO SOLVE.";
212   INPUT S
225   LET N = 0
230   PRINT "CHOOSE THE RANGE OF DIFFICULTY."
235   PRINT  TAB( 10) "EASY (1-10) - TYPE E"
240   PRINT  TAB( 10) "MEDIUM (11-30) - TYPE M"
245   PRINT  TAB( 10) "HARD (31-100) - TYPE H"
250   INPUT D$
255   IF D$ = "E" THEN  LET A = 10:B = 1
256   IF D$ = "M" THEN  LET A = 30:B = 11
257   IF D$ = "H" THEN  LET A = 100:B = 31
259   REM WILL COUNT NUMBER OF PROBLEMS
260   FOR C = 1 TO S STEP 1
265   REM
266   REM  THE WORKING ADDITION
270   LET X =  INT (A * RND (1)) + B
275   LET Y =  INT (A * RND (1)) + B
```

```
280     LET Z = X + Y
285     PRINT "WHAT IS "X"+"Y
290     INPUT G
300     IF G = Z THEN  GOTO 330
310     PRINT "SORRY! THE ANSWER IS ";Z: NEXT C
320     GOTO 345
330     PRINT "YOU GOT IT!!"
331     PRINT
335     LET N = N + 1
340     NEXT C
345     PRINT "YOU GOT "N" OUT OF "S" CORRECT"
350     PRINT "WOULD YOU LIKE MORE ADDITION
        PROBLEMS?"
355     PRINT "(Y OR N)": INPUT T$
360     IF T$ = "Y" THEN GOTO 210
370     RETURN
400     REM  PROGRAM WILL RUN SUBTRACTION PROBLEMS
405     REM
410     PRINT "SELECT THE NUMBER OF PROBLEMS YOU"
411     PRINT "WOULD LIKE TO SOLVE";
420     INPUT S
425     LET N = 0
430     PRINT "CHOOSE THE RANGE OF DIFFICULTY"
435     PRINT  TAB( 10) "EASY (1-10)-TYPE E"
440     PRINT  TAB( 10) "MEDIUM (11-30) - TYPE M"
445     PRINT  TAB( 10) "HARD (31-100) - TYPE H"
450     INPUT D$
455     IF D$ = "E" THEN  LET A = 10:B = 1
456     IF D$ = "M" THEN  LET A = 30:B = 11
457     IF D$ = "H" THEN  LET A = 100:B = 31
459     REM WILL COUNT NUMBER OF PROBLEMS
460     FOR C = 1 TO S STEP 1
465     REM
466     REM  THE WORKING SUBTRACTION
470     LET X =  INT (A * RND (1)) + B
475     LET Y =  INT (A * RND (1)) + B
480     LET Z = X - Y
485     PRINT "WHAT IS "X"-"Y
490     INPUT G
500     IF G = Z THEN  GOTO 530
510     PRINT "SORRY! THE ANSWER WAS";Z: NEXT C
520     GOTO 545
530     PRINT "YOU GOT IT!!"
535     LET N = N + 1
540     NEXT C
545     PRINT "YOU GOT "N" OUT OF "S" CORRECT"
550     PRINT "WOULD YOU LIKE MORE SUBTRACTION"
552     PRINT "PROBLEMS? (Y OR N)"
555     INPUT T$
560     IF T$ = "Y" THEN GOTO 410
570     RETURN
600     REM PROGRAM WILL RUN MULTIPLICATION PROBLEMS
605     REM
610     PRINT "SELECT THE NUMBER OF PROBLEMS YOU
        WOULD"
612     PRINT "LIKE TO SOLVE.";
620     INPUT S
625     LET N = 0
630     PRINT "CHOOSE THE RANGE OF DIFFICULTY."
635     PRINT  TAB( 10)"EASY (1-10) - TYPE E"
640     PRINT  TAB( 10)"MEDIUM (11-30) - TYPE M"
```

```
645    PRINT  TAB( 10)"HARD (31-100) - TYPE H"
650    INPUT D$
655    IF D$ = "E" THEN  LET A = 10:B = 1
656    IF D$ = "M" THEN  LET A = 30:B = 11
659    REM   WILL COUNT NUMBER OF PROBLEMS
660    FOR C = 1 TO S STEP 1
665    REM
666    REM   THE WORKING MULTIPLICATION
667    REM
670    LET X =   INT (A *  RND (1)) + B
675    LET Y =   INT (A *  RND (1)) + B
680    LET Z = X * Y
685    PRINT "WHAT IS "X"*"Y
690    INPUT G
700    IF G = Z THEN  GOTO 730
710    PRINT "SORRY! THE ANSWER IS";Z: NEXT C
720    GOTO 745
730    PRINT "YOU GOT IT!!"
731    PRINT
735    LET N = N + 1
740    NEXT C
745    PRINT "YOU GOT "N" OUT OF "S" CORRECT"
750    PRINT "WOULD YOU LIKE MORE MULTIPLICATION"
751    PRINT "PROBLEMS? (Y OR N)": INPUT T$
760    IF T$ = "Y" THEN  GOTO 610
770    RETURN
800    REM PROGRAM WILL RUN DIVISION PROBLEMS
805    REM
806    REM
810    PRINT "SELECT THE NUMBER OF PROBLEMS YOU"
811    PRINT "WOULD LIKE TO SOLVE";
812    INPUT S
825    LET N = 0
830    PRINT "CHOOSE THE RANGE OF DIFFICULTY"
835    PRINT  TAB( 10)"EASY(1-9) - TYPE E"
840    PRINT  TAB( 10)"HARD(10-30) - TYPE H"
850    INPUT D$
855    IF D$ = "E" THEN  LET A = 9:B = 1
856    IF D$ = "H" THEN LET A = 21:B = 10
857    REM WILL COUNT NUMBER OF PROBLEMS
860    FOR C = 1 TO S STEP 1
865    REM
866    REM  THE WORKING DIVISION
870    LET X =   INT (9 *  RND (1)) + 1
875    LET Y =   INT (A *  RND (1)) + B
876    LET V = X * Y
880    LET Z = V / X
885    PRINT "WHAT IS "V"/"X
890    INPUT G
900    IF G = Z THEN  GOTO 930
910    PRINT "SORRY! THE ANSWER IS ";Z: NEXT C
920    GOTO 945
930    PRINT "YOU GOT IT!!"
931    PRINT
935    LET N = N + 1
940    NEXT C
945    PRINT "YOU GOT "N" OUT OF "S" CORRECT"
950    PRINT "WOULD YOU LIKE MORE DIVISION
       PROBLEMS?"
955    PRINT "(Y OR N)": INPUT T$
960    IF T$ = "Y" THEN  GOTO 810
970    RETURN
```

In addition to the preceding program, we can use the computer to begin the study of examples, exercises, and/or problems in word form, as the following example indicates.

```
10   REM  CREATING WORD EXERCISES FOR ADDITION
15   PRINT "TODAY YOU ARE GOING TO WRITE YOUR WORD
     EXERCISES"
20   PRINT "WRITE THE NAME OF A FRIEND"
25   INPUT F$
30   PRINT "TYPE A NUMBER BETWEEN 1 AND 10"
35   INPUT N
40   PRINT "WRITE THE NAME OF YOUR FAVORITE ANIMAL"
45   INPUT A$
50   PRINT "WRITE THE NAME OF ANOTHER FRIEND"
52   PRINT
55   INPUT FF$
60   PRINT "TYPE ANOTHER NUMBER BETWEEN 1 AND 10"
65   INPUT P
70   PRINT "WRITE THE NAME OF ANOTHER ANIMAL"
75   INPUT AA$
77   PRINT
80   PRINT F$;" SAW ";N;" ";A$;"S"
85   PRINT
90   PRINT FF$;" SAW ";P;" ";AA$;"S"
95   PRINT
100  PRINT "HOW MANY ANIMALS DID ";F$;" AND
     ";FF$;" SEE?"
105  PRINT
110  INPUT Q
120  IF Q = N + P THEN  GOTO 150
130  IF Q < > N + P THEN  GOTO 140
140  PRINT "LET'S TRY AGAIN"
145  GOTO 80
150  PRINT "GOOD JOB, DO YOU WISH TO TRY AGAIN?"
160  PRINT "TYPE YES IF YOU WANT TO TRY AGAIN AND
     NO IF YOU DO NOT WISH TO TRY AGAIN"
170  INPUT D$
180  IF D$ = "YES" THEN  GOTO 20
190  IF D$ = "NO" THEN  GOTO 200
200  PRINT "IT WAS NICE TO WORK WITH YOU"
210  END
```

These examples of the use of multimedia materials or instructional aids are only introductory in nature and only scratch the surface in this area. They represent (in mathematical terms) a subset of the domain of providing experiences for the mildly handicapped, which can enable them to function successfully in the world of technology.

As Edward Caine (1984, 239) has observed, "technology, properly developed, can overcome physical barriers, facilitate communication, compensate for biological deficiencies, and serve as daily living prosthetic devices." We have a responsibility to develop new instructional strategies that maximize the potential of technology in the education of exceptional children.

REFERENCES

Ackerman, P. Y., Anhalt, J. M., and Dykman, R. S. 1986. Arithmetic automatization failure in children with attentional and reading disorders: Associations and sequels. *Journal of Learning Disabilities 19:* 222–232.

Allardice, B., and Ginsburg, H. 1983. Children's psychological difficulties in mathematics. In *The development of mathematical thinking,* ed. H. Ginsburg. New York: Academic Press.

Alley, B., and Deshler, D. D. 1979. *Teaching the learning disabled adolescent: Strategies and methods.* Denver, Colo.: Love Publishing Co.

Ashlock, R. 1982. *Error patterns in computation.* Columbus, Ohio: Charles Merrill.

Ballew, H., and Cunningham, J. 1983. Diagnosing strengths and weaknesses of sixth-grade students in solving work problems. *Journal for Research in Mathematics Education 13:* 202–210.

Blankenship, C. S. 1985. Linking assessment to curriculum and instruction. In *Practical mathematics appraisal of the learning disabled,* ed. J. F. Cawley, 59–81. Rockville, Md.: Aspen Systems.

Blankenship, C. S. 1984. Curriculum and instruction: An examination of models in special and regular education. In *Developmental teaching of mathematics for the learning disabled,* ed. J. F. Cawley, 29–54. Rockville, Md.: Aspen Systems.

Blankenship, C. S. 1982. Programming generalization of computational skills. *Learning Disability Quarterly 5:* 152–162.

Blankenship, C. S., and Lovitt, T. C. 1976. Story problems: Merely confusing or downright befuddling? *Journal of Research in Mathematics Education 7:* 290–298.

Bloom, B., ed. 1956. *Taxonomy of educational objectives.* New York: D. McKay.

Breault, M. A. 1983. *The effects of extraneous information on word problems.* Unpublished doctoral dissertation, University of Connecticut.

Brigance, A. 1977. *Brigance diagnostic inventory of basic skills.* Woburn, Mass.: Curriculum Associates.

Bruner, J. S. 1963. *Toward a theory of instruction.* Cambridge, Mass.: Harvard University Press.

Buswell, G. T., and John, L. 1925. *Diagnostic chart for fundamental processes in arithmetic.* Indianapolis, Ind.: Bobbs-Merrill.

Caine, E. J., Jr. 1984, Summer. The challenge of technology: Educating the exceptional child for the world of tomorrow. *Teaching Exceptional Children,* 239–241.

Cawley, J. F., Norlander, K., Bates, H., and Sayre, J. In press. *Learner activity program: Word problems.* Rockville, Md.: Aspen Systems.

Cawley, J. F., Fitzmaurice-Hayes, A. M., Shaw, R. A., Norlander, K., and Sayre, J. 1987. *Learner activity program: Developmental mathematics.* Rockville, Md.: Aspen Systems.

Cawley, J. F., and Miller, J. H. 1986. Appraisal and therapy in mathematics. *The Educational Therapist 7:* 2–5.

Cawley, J. F. 1985a. Mathematics and vocational preparation for the learning disabled. In *Secondary school mathematics for the learning disabled,* ed. J. F. Cawley, 201–234. Rockville, Md.: Aspen Systems.

Cawley, J. F. 1985b. Learning disability and mathematics appraisal. In *Practical mathematics appraisal of the learning disabled,* ed. J. F. Cawley, 1–41. Rockville, Md.: Aspen Systems.

Cawley, J. F. 1984a. Learning disabilities: Issues and alternatives. In *Developmental teaching of mathematics for the learning disabled,* ed. J. F. Cawley, 1–28. Rockville Md.: Aspen Systems.

Cawley, J. F. 1984b. An integrative approach to needs of learning disabled: Expanded use of mathematics. In *Developmental teaching of mathematics for the learning disabled,* ed. J. F. Cawley, 81–94. Rockville, Md.: Aspen Systems.

Cawley, J. F. 1984c. Selection, adaptation, and development of curricula and instructional materials. In *Developmental teaching of mathematics for the learning disabled,* ed. J. F. Cawley, 227–252. Rockville, Md.: Aspen Systems.

Cawley, J. F., Fitzmaurice, A. M., Shaw, R. A., and Norlander, K. 1980. *Multi-modal-mathematics.* Unpublished experimental curriculum, University of Connecticut.

Cawley, J. F., Fitzmaurice, A. M., Shaw, R. A., Kahn, H., and Bates, H. 1979a. LD youth and mathematics: A review of characteristics. *Learning Disability Quarterly 2:* 25–41.

Cawley, J. F., Fitzmaurice, A. M., Shaw, R. A., Kahn, H., and Bates, H. 1979b. Math word problems: Suggestions for LD students. *Learning Disability Quarterly 2:* 25–41.

Cawley, J. F., Fitzmaurice, A. M., Goodstein, H. A., Lepore, A., Sedlak, R., and Althaus, V. 1974–1976. *Project MATH.* Tulsa, Okla.: Educational Progress.

Cawley, J. F., and Goodman, J. O. 1968. Interrelationships among mental abilities, reading, language arts and arithmetic with the mentally handicapped. *Arithmetic Teacher 15:* 631–636.

Cawley, J. F., and Goodman, J. O. 1969. Arithmetical problem solving: A program demonstration by teachers of the mentally retarded. *Exceptional Children 36:* 83–90.

Cohen, S. A., and Stover, G. 1981. Effects of teaching sixth-grade students to modify format variables of math word problems. *Reading Research Quarterly 16:* 175–200.

Connolly, A., Nachtman, W., and Prichett, E. 1971. *Key Math Diagnostic Arithmetic Test.* Circle Pines, Minn.: American Guidance Service.

Cruickshank, W. C. 1948. Arithmetic ability of mentally retarded children. 1: Ability to differentiate extraneous materials from needed arithmetic facts. *Journal of Educational Research 42:* 161–170.

Davis, R. 1984. *Learning mathematics.* Norwood, N.J.: Ablex Publishing.

Executive Summary 1984. Sixth annual report to Congress on the implementation of Public Law 94–142: The education of all handicapped children's act. *Exceptional Children 51:* 199–202.

Fitzmaurice-Hayes, A. M. 1985. Assessment of the severely impaired mathematics student. In *Practical mathematics appraisal of the learning disabled,* ed. J. F. Cawley, 249–258. Rockville, Md.: Aspen Systems.

Fleischner, J., and O'Loughlin, M. 1985. Solving story problems: Implications of research for teaching the learning disabled. In *Cognitive Strategies and Mathematics for the Learning Disabled.* ed. J. F. Cawley, 163–182. Rockville, Md.: Aspen Systems.

Fleischner, J., Garnett, K., and Shepherd, M. 1982. *Proficiency in arithmetic basic fact computation of learning disabled and non-learning disabled children.* (Tech. Rep. No. 9). New York: Teachers College, Columbia University, Research Institute for the Study of Learning Disabilities.

Gelman, R., and Gallistel, R. C. 1986. *The child's understanding of number.* Cambridge, Mass.: Harvard University Press.

Ginsburg, H. 1986. *Children's arithmetic: How they learn it and how you teach it.* Austin, Tex.: ProEd.

Goodstein, H. A. 1974. Individualizing instruction through matrix teaching. *Education and Training of the Mentally Retarded 9:* 189–190.

Goodstein, H. A., and Sedlak, R. 1974. *The role of memory in the verbal problem solving of average, mentally retarded and learning disabled children.* Unpublished research paper, University of Connecticut.

Grise, P. J. 1980. Florida's minimum competency testing program for handicapped students. *Exceptional Children 47:* 186–193.

Hewett, F. M. 1968. *The emotionally disturbed child in the classroom.* Boston: Allyn & Bacon.

Jastak, J. F., and Jastak, S. R. 1965. *Wide range achievement tests: Arithmetic.* Bloomington, Del.: Guidance Associates.

Kosc, L. 1974. Developmental dyscalculia. *Journal of Learning Disabilities 7:* 164–177.

Kraner, R. E. 1976. *Kraner preschool math inventory.* Austin, Tex.: Learning Concepts.

Krupski, A. 1985. Variations in attention as a function of classroom task demands in learning handicapped and CA matched nonlearning handicapped children. *Exceptional Children 52:* 52–56.

Lancy, D. F. 1983. *Cross cultural studies in cognition and mathematics.* New York: Academic Press.

Levine, H., and Langress, L. 1985. Everyday cognition among mildly retarded adults: An ethnographic approach. *American Journal of Mental Deficiency 90:* 18–26.

Lyon, G. R. 1985. Identification and remediation of learning disability subtypes: Preliminary findings. *Learning Disabilities Focus 1:* 21–35.

McCracken, G., and Walcutt, C. 1963. *Basic Reading* (Teacher ed.). Philadelphia: Lippincott.

McKinney, J., and Haskins, K. 1980. *Performance of exceptional students on the North Carolina Minimum Competency Test. Final Report.* Chapel Hill, N.C.: University of North Carolina.

McLeod, T., and Armstrong, S. 1982. Learning disabilities in mathematics: Skill deficits and remedial approaches at the intermediate and secondary level. *Learning Disability Quarterly 5:* 305–311.

Menninger, K. 1977. *Number words and number symbols: A cultural history of numbers.* Cambridge, Mass.: M.I.T. Press.

Meyen, E. 1968. *An investigation of age placement, difficulty and importance of basic skills in the curriculum of EMR students.* Unpublished doctoral dissertation, University of Iowa.

Montessori, M. 1917. *Advanced Montessori method.* Reprint. Cambridge, Mass.: Robert Bentley, 1964.

Morse, W. 1980. Worksheet on life-spaced interviewing for teachers. In *Conflict in the classroom,* ed. N. Long, W. Morse, and R. Newman. Belmont, Calif.: Wadsworth.

National Assessment of Educational Progress. 1974. Washington, D.C.: U.S. Dept. of Education.

National Council of Teachers of Mathematics. 1973. *Instructional aids in mathematics.* Washington, D.C.: author.

National Council of Teachers of Mathematics. 1980. *Priorities in School Mathematics.* Reston, VA.: author.

Norman, C.A., and Zigmond, N. 1980. Characteristics of children labeled and served as learning disabled in school systems affiliated with child service demonstration centers. *Journal of Learning Disabilities 13:* 542–548.

Penner, W. 1972. *Effect of cue word form class on solving of arithmetic word problems by the mentally handicapped.* Unpublished doctoral dissertation, University of Connecticut.

Piaget, J. 1959. *The intelligence of children.* New York: International Universities Press.

Redl, F. 1959. The concept of life-space interview. *American Journal of Orthopsychiatry 29:* 1–18.

Reisman, F. K. 1978. *A guide to the diagnostic teaching of arithmetic.* Columbus, Ohio: Charles E. Merrill.

Reisman, F. K., and Kauffman, S. H. 1980. *Teaching mathematics to children with special needs.* Columbus, Ohio: Charles E. Merrill.

Resnick, L., and Resnick, D. 1985. Standards, curriculum, and performance: A historical and comparative perspective. *Educational Researcher 14:* 5–29.

Resnick, L., and Ford, W. 1981. *The psychology of mathematics for instruction.* Hillsdale, N.J.: Lawrence Erlbaum Associates.

Resnick, L. 1983. A developmental theory of number understanding. In *The development of mathematical thinking,* ed. H. Ginsburg, 110–152. New York: Academic Press.

Rice, W., Guskey, T., Pearlman, C., and Rice, M. 1980. *Minimum essentials test.* Glenview, Ill.: Scott, Foresman.

Schenck, W. 1973. Pictures and the indefinite quantifier in verbal problem solving among EMR children. *American Journal of Mental Deficiency, 78:* 272–276.

Schoenfeld, A. H. 1982. Some thoughts on problem solving research and mathematics education. In *Mathematical problem solving: Issues in research,* ed. F. Lester and J. Garofalo. Philadelphia: Franklin Institute.

Sedlak, R. 1973. *Performance of good and poor problem solvers on arithmetic problems presented in modified cloze format.* Unpublished doctoral dissertation, Pennsylvania State University.

Seguin, E. 1849. *Idiocy and its treatment.* Educational Reprint. New York: Teachers College, Columbia University, 1907.

Shaw, R. A. 1985. Adapting the college preparatory program for the learning disabled. In *Secondary school mathematics for the learning disabled,* ed. J. F. Cawley, 271–290. Rockville, Md.: Aspen Systems.

Shaw, R. A. 1981. Diagnosing and using nonword problems as aids to thinking and comprehension. *Topics in Learning and Learning Disabilities 1:* 11–29.

Sorensen, D. 1984. *Number stumper*. Menlo Park, Calif.: Learning Co.

Strauss, A. A., and Lehtinen, L. E. 1947. *Psychopathology and education of brain-injured children*. New York: Grune & Stratton.

Thornton, C. A., and Toohey, M. A. 1985. Basic math facts: guidelines for teaching and learning. *Learning Disabilities Focus 1:* 44–57.

Torgenson, J. 1982. The learning disabled as an inactive learner: Educational implication. *Topics in Learning and Learning Disabilities 2:* 45–52.

Trenholme, B., Larsen, S. C., and Parker, R. 1978. The effects of syntactic complexity upon arithmetic performance. *Learning Disability Quarterly 1:* 80–85.

Troutman, A. 1980. *Diagnosis: An instruction aid: Mathematics*. Chicago: Science Research Associates.

Truesdell, L. 1985. Making it in the mainstream: Special education student behavior and academic success. Paper presented at the Annual Meeting of the American Educational Research Association. Chicago, Ill.

Van de Walle, J., and Thompson, C. A. 1980. Let's do it. Fractions with counters. *The Arithmetic Teacher 28* (2): 6–11.

Warner, M., Alley, G., Schumaker, J., Deshler, D., and Clarke, F. 1980. *An epidemiological study of learning disabled adolescents in secondary schools: Achievement and ability, socioeconomic and school experiences*. Research Report 13, Institute for Research in Learning Disabilities, University of Kansas, Lawrence.

Webster, R. E. 1981. *Learning Efficiency Test*. Novato, Calif.: Academic Therapy Publications.

Webster, R. E. 1980. Short-term memory in mathematics proficient and mathematics deficient students as a function of input-output modality pairings. *The Journal of Special Education 14:* 67–78.

Zigler, E., Balla, and Hodapp. 1984. On definition and classification of mental retardation. *American Journal on Mental Deficiency 89:* 215–230.

INDEX